Nathan lives on a small farm in a quaint village outside of York. He wrote *The History of Life on Earth* while working on the family farm over the summer vacations from university. Nathan studied chemistry at the University of York and has a keen interest in all aspects of science. Indulging his childhood intrigue of prehistoric life, Nathan compiled his knowledge of the subject into his first book. The objective of this was to fulfil his passion for writing and engage the reading audience with the fascinating story of evolution and share his enthusiasm for science. His hopes and aspirations for the future are to continue to write about science and inspire his readers.

Nathan Keighley

THE HISTORY OF LIFE ON EARTH

THE STORY OF EVOLUTION ON OUR PLANET

AUSTIN MACAULEY PUBLISHERS™

LONDON • CAMBRIDGE • NEW YORK • SHARJAH

A CIP catalogue record for this title is available from the British Library.

ISBN 9781528903028 (Paperback)
ISBN 9781528903035 (E-Book)

www.austinmacauley.com

First Published (2018)
Austin Macauley Publishers Ltd™
25 Canada Square
Canary Wharf
London
E14 5LQ

Table of Content

Preface

The History of Life on Earth is designed for a wide audience; essentially anyone who is intrigued to learn about prehistoric life will hopefully find it of interest. There are no academic prerequisites to understanding the material because the scientific background is well explained, but readers with enthusiasm for science would probably enjoy the text most. The content of the manuscript is based on the universally acknowledged fundamental and well-understood principles of biology and palaeontology that can be found in any broad-spectrum textbooks, but written in my own words as an engaging narrative; designed to be a refreshing read in continuous prose, rather than a systematic reference book.

The style of the manuscript differs from many books in this genre, being incredibly concise. I have written it in the format of a series of essays, which constitute the sub-chapters of the broader topic chapters, where each chapter is in continuity with the next. Each chapter is creatively written to offer a succinct description of fundamental evolutionary principles and description of prehistoric life. My aspiration is to produce a text that is exciting to read by offering an imaginatively written narrative, covering the entire history of life in a brief, but inventive, manuscript. This is in contrast to typical texts, which are much more lengthy, but not necessarily more rigorous. The objective of my manuscript is to provide a short, but detailed account of the history of life on Earth, so that the reader can readily learn the essentials of the subject and retain the important information without distraction from lengthy preambles and summaries. In conclusion, my goal is to convey my passion for a subject in which I am not an expert, but have a keen interest.

Introduction

The earliest aeon of the Earth is the Hadean, which began 4.567 billion years ago, when the molten Earth was formed. As the core and mantle separated, and the surface cooled to form the crust, considerable volcanic activity and degassing occurred, producing an atmosphere composed of nitrogen, carbon dioxide, methane, ammonia, hydrogen sulphide and water vapour, which eventually condensed to form oceans. Conditions were set for the creation of life approximately 3,500 million years ago, during the Archean, when the atmospheric constituents, activated by electrical energy from lightning, reacted to form amino acids and the genesis of life began. With the essential molecules for life generated, the first cells formed; bacterial prokaryotic microbes developed, living in colonies called stromatolites. They could photosynthesise, taking in carbon dioxide and releasing oxygen into a hitherto anoxic atmosphere.

Plentiful supplies of oxygen stimulated the development of eukaryotic cells in the Proterozoic and, by 600 Ma, multicellular organisms developed known as the Ediacaran faunas and floras. By the beginning of the Palaeozoic, in the Cambrian, animals were capable of extracting calcium carbonate from the surrounding seawater to produce shells. This resulted in an evolutionary arms race, propagating the proliferation of invertebrates, which dominated in the Palaeozoic. Vertebrates, which evolved in the Ordovician, by the Devonian, undertook a major evolutionary step in the history of life on Earth: the conquest of land. This provided an abundance of opportunities for the diversification of the vertebrates; enabling them to develop into the rich variety of animals observed in the geological record.

Eon	Era	Period	Age (Ma)
Phanerozoic	Cainozoic	Quaternary Tertiary	2.6–0 65–2.6
Phanerozoic	Mesozoic	Cretaceous Jurassic Triassic	145–65 200–145 250–200
Phanerozoic	Palaeozoic	Permian Carboniferous Devonian Silurian Ordovician Cambrian	300–250 360–300 420–360 445–420 490–445 540–445
Proterozoic	Ediacaran		635–540 2,500–540
Archean			4,000–2,500
Hadean			4,567–4,000

Figure 1: The geological column, showing the ages in the history of the Earth.

The history of life preserved in rocks is essentially a document recording the major evolutionary processes undertaken by life on Earth. While evolution is underpinned by molecular genetics, its mechanical application to life can be described through Darwin's theory of natural selection. The general idea behind the theory is straightforward: species reproduce more rapidly than is need for them to maintain their numbers, yet populations remain stable due to the competition that exists within and between species for food, space etc. Variation exists within a species, and individuals with the favourable characteristics survive the intense competition and their offspring inherit the favourable characteristics.

Variation comes from DNA: the genetic material that is inherited by the offspring from their parents. DNA is essentially the instruction manual for making proteins and building a body;

hence, variations in its sequence can effect an organism's phenotype (appearance). The region on a DNA molecule that codes for a particular protein is called a gene, which may exist in different forms called alleles. Variation in a population is linked with the number of alleles present. Variation occurs as cells divide during mitosis and meiosis, when chromosomes are reshuffled. However, a more radical source of variation comes from mutation. Spontaneous changes in the DNA sequence of an organism can lead to a discrete change in phenotype, which may or may not be advantageous. A given species is often geographically isolated into separate populations. When the spread of mutation is independent of a species as a whole, eventually the individual populations may vary genetically to the extent that they can no longer interbreed: a new species is formed (this is called speciation).

Throughout the fossil record, it is evident that different modes of evolution exist. Evolutionary processes, interpreted from the observation of the descent of evolving lineages, can be considered as two main modes: microevolution and macroevolution. The changing interactions between organisms and their environment, leading to speciation, can be attributed to microevolution. The steady process of natural selection envisaged by Darwin is termed phyletic gradualism. However, these periods of stasis end as species become extinct during a catastrophic event, leading to a punctuated equilibrium, which provides an opportunity for the abrupt development of new species relative to the time spent during stasis. Macroevolution deals with these larger scale changes. This provides an overview of the history of life on Earth.

Throughout the course of evolution, through gradual change and intermittent periods of rapid propagation of new species, higher-level taxonomic groups are derived. In order to catalogue the variety of faunas and floras that exist and have existed on planet Earth, a rigorous classification system is required. Organisms are classified in accordance with a hierarchy of taxonomic levels.

```
Kingdom              (prefixes) super-
  Phylum                      sub-
    Class                    infra-
      Order
        Family
          Genus
            Species
```

Figure 2: Linnean system of classification; the taxonomic groups.

Using this method of classification, groups of organisms may be arranged diagrammatically into "family trees" called cladograms, which can be used to demonstrate the relationships between different organisms. On a more specific level, the scientific names designated to organisms are written in italics, with the first letter of the name of the genus in uppercase and the species name in lowercase. This method of classification, in some instances, can be misleading though, because the specific name of the species may be counterintuitive when compared to common names that are frequently used. For example, *Puffinus puffinus* is not a puffin, but a shearwater, and *Pinguinus* isn't a penguin; it is the extinct great auk. The purpose of this system, invented by Swedish biologist Carl Linnaeus, is to provide an unambiguous classification of organisms, with a unique name that is standardised among scientific communities around the world.

Cladistics is a useful approach to taxonomy because it traces the evolutionary lineages by which species appear and use that phylogeny to organise species into clades, that is, groups of species that descended from one ancestral species, which is represented as a branch on the cladogram. Thus cladograms visually depict relationships between organisms: clades of organisms exist in a hierarchy of scales, where the terminal branches may represent single species and the confluences represent higher-level taxonomic ranks, but each branching event is due to one ancestral species.

The diagrams are arranged by identifying groups of species with certain characteristics in common that were evolved as newly derived features in a common ancestor; the primitive state.

11

Take birds for example; all birds have feathers, which is a derived feature unique to that class of animals. Perhaps a feathered skin evolved from one covered with scales; the primitive state, which is the source of the clade. The hypothesised phylogeny can be further supported by examination of other common features, such as the wishbone, breastbone and a box-like pelvis, present in all birds. However, complications can arise when a derived feature is present in different species outside a clade. Bats and birds both have wings, a derived feature modified from some other structure, but have very few other derived features in common. This is an example of parallel evolution, where the same derived feature is evolved independently. Hence, cladograms are drawn as a hypothesis deduced as shared primitive characteristics are separated from discrete derived features. Darwin's postulate of modification with descent explains how advances from a primitive ancestor in different directions may produce several new species. This, essentially, is the logic used in cladistics.

Counterintuitive patterns can arise in cladistics. Inequalities may be present between cladistics status and taxonomic rank. For example, it is misleading to consider all the classes of vertebrates as being equal in rank because fish, amphibians, reptiles, birds and mammals are not derived separately. While fish ultimately descended into living forms, they also derived the clade of tetrapods, which include amphibians and reptiles. Birds and mammals are then equally clades of derived reptiles. The cladogram conveniently diagrammatically portrays these classifications.

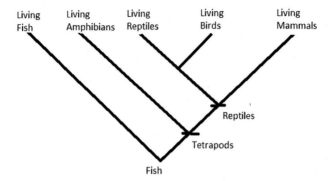

Figure 3: Cladogram showing the evolutionary relationships between all the classes of vertebrates.

Cellular Beginnings
Origin of the Cell

All living entities are made of cells. While complex organisms are cellular cities, simple life forms are solitary cells, and this is the format on which life is thought to have begun. The origin of the cell, and indeed the origin of life, is believed to have occurred 3.5 billion years ago, in the Earth's early history when the environment was violently volcanic, following the planet's formation. Somewhere, perhaps "in a little pond," postulated Charles Darwin, life started: a living system enclosed in a fatty capsule, which possessed a genetic code inscribed on RNA, with which it could build proteins and reproduce itself. This was the first cell.

The chemical and physical environment of the early Earth provided the unique conditions necessary for the creation of life. When the Earth was newly formed, violent volcanic activity at the surface supplied an atmosphere composed of carbon dioxide, methane, ammonia, hydrogen sulphide and water vapour. The lack of ozone allowed intense UV radiation through the atmosphere which, along with the electrical discharge from lightning, provided the energy needed for gaseous molecules to react to form simple organic molecules, such as amino acids and nucleotides, which are vital for life. This process was replicated experimentally in the laboratory by Stanley Miller and Harold Urey in 1952. A cocktail of organic molecules formed from the mixture of gases that were dissolved in water, when exposed to UV radiation and electrical discharge. While profound in regards to its biological implications, this discovery exemplifies the fundamental processes that occur within a chemical system. An outside source of energy (the UV radiation and electricity) activated the molecules, which were able to interact and react because of their enhanced mobility within the aqueous medium. The water, being chemically inert, also provided protection to the reaction system.

This process may have occurred in water droplets in the early atmosphere and later in the oceans and pools formed as the water droplets condensed and fell as rain, just as Darwin had speculatively suggested. The necessity for biological reactions to occur in an aqueous medium persists within the body of cells today. One crucial biological process is the synthesis of macromolecules; as monomers, such as amino acids and nucleotides, associate to form polymers. Crucially, this process must have occurred in the primordial oceans. Such an environment is far from chemical equilibrium; these conditions involved polymers influencing subsequent reactions, acting in an autocatalytic system that would have properties characteristic of living matter: it would comprise a non-random selection of interacting molecules; it would tend to reproduce itself; it would compete with other systems dependent on the same feedstock; and, if deprived of the required feedstock or placed under unfavourable conditions, will decay towards equilibrium and die.

Polynucleotides undergo complementary base pairing, where one acts as the template for the formation of another from constituent nucleotide monomers. RNA (ribonucleic acid) is formed from long sequences of polynucleotides that consist of only four bases: adenine, uracil, cytosine and guanine. RNA can form a unique three-dimensional shape, when electrostatic bonds form between different sections of the chain. This folded structure governs the molecules' stability, reaction with other molecules, and its ability to replicate, so not all polynucleotide shapes will be equally successful in the replicating mixture. Moreover, errors that may inevitably occur during the replication process produce new variations of RNA that replicate themselves and a type of natural selection occurs, where favourable sequences of nucleotides, which determine a particular folded structure, predominate under certain conditions.

RNA essentially has two properties that are essential for evolution: it can carry information enclosed in the nucleotide sequence, analogous to the organism's genotype, and its folded structure enables it to interact selectively with other molecules, namely amino acids, when proteins are built to form a body; the overall appearance (phenotype) is created. It is therefore speculated that around 3.5–4 billion years ago, replicating systems of RNA started the evolutionary process.

In order to proceed in this manner, RNA had to cooperate mutually with polypeptides; as their catalytic abilities are far superior. Polypeptides are combined sequences of amino acids, which could provide an alternative route for the replication of RNA that has a lower required energy threshold, making the process more efficient. Meanwhile, RNA molecules with the ability to bind with amino acids could group together and manufacture polypeptides from their constituent amino acids. In summary, any polynucleotide that could promote the synthesis of a polypeptide with the ability to mechanically provide a more efficient route for the replication of that polynucleotide would be at a selective advantage. Today, RNA provides a vital role in the cell for the manufacture of long polypeptides: proteins.

Perhaps the most crucial event leading to the formation of the first cell must have been the development of the outer membrane. An enclosed membrane could contain the RNA and assorted polypeptides in an isolated environment devoid of competing varieties of RNA, so that only indigenous varieties could benefit from the catalytic polypeptides. The cell membrane likely developed as a consequence of the chemical nature of the phospholipids of which it consists. Phospholipids are like fats and oils; they are described as being amphipathic, having a hydrophilic head and a water-repellent tail, where the homologous parts of the molecules contact each other as the phospholipids aggregate in aqueous solution to form bilayers. Such bilayers not only form a skin on the surface of the water, but also assemble as vesicles whose aqueous contents are isolated from the external medium.

These primitive cells are believed to have carried their genetic information on RNA. Eventually though, RNA was replaced by DNA (deoxyribonucleic acid), consisting of the same nucleotides, except for uracil, which is replaced by thymine. The notorious double helix meant that DNA is much more stable than RNA, therefore much less susceptible to corruption from external chemical sources. Consequently, incredibly long strands of DNA could develop, capable of storing and transmitting much greater quantities of genetic information, enabling cells to become increasingly complex.

From Prokaryotes to Eukaryotes

Because of the similarities in metabolic pathways and DNA sequences between organisms, it is thought that all life is derived from one type of cell. This "original cell" has become known as the last universal common ancestor (LUCA); thought to have come into existence 3.5 billion years ago. No doubt these early cells were fundamentally basic; probably similar to the unicellular organisms to which they are ancestral: the Prokaryotes. Prokaryotes are simplistic single-celled organisms, yet they exhibit evolutionary complexities in their habits. The early cells of the Proterozoic were almost certainly prokaryotic, but would have lacked many of the advances of their modern forms. With the possession of DNA and resultant evolutionary potential, early prokaryotes were able to evolve, diversify and adapt to the volatile environment of the early Earth.

The early prokaryotic bacterial cells, which had no obvious internal structures, were protected by a strong cell wall, just like bacteria today. The cell membrane is selectively permeable to allow the passage and assimilation of nutrients. Cyanobacteria, or blue-green algae, have been preserved in stromatolite formations from pre-Cambrian times; they have no nuclei; their DNA is contained loosely in the cytoplasm. The ability of organisms to replicate their DNA enables them to reproduce. Mutations occur during DNA replication; an error in the process may lead to an alteration in the base sequence and ultimately a different protein is produced, which is inherited by the next generation. The incumbent mutagenic protein translated from the new DNA will often be defective and therefore impede on the cell's proliferation. However, purely by chance and with a small probability, it may be the case that the novel protein is advantageous to the cell, perhaps enabling the metabolism of more nutrients. The small bacterial cells can readily replicate by dividing into two during binary fission, so they can rapidly reproduce. This property means that the occurrence of spontaneous mutations in a fast-growing population accelerates natural selection, enabling bacteria to respond quickly to changing environmental conditions. Hence, bacteria can inhabit a variety of ecological niches, including hostile "extremophile" locations, as they can develop new metabolic pathways to cope with the change.

By developing new metabolic pathways, that is, novel ways to utilise the available resources, bacteria could adapt to exploit a variety of carbon sources; even a single source, such as glucose exclusively, in order to derive chemical energy and synthesise every type of organic molecule that the cell requires. As populations grew, competition increased among the early cells. It became a necessity to improve metabolic reactions. A major step in evolution was to use enzymes, which are proteins with a specific shape, to catalyse biological reactions. Those cells that could use the natural resources faster and more efficiently had the selective advantage, depriving other cells of these resources.

One of the earliest metabolic pathways must have been glycolysis; the first stage of respiration, where glucose is degraded in the absence of oxygen to produce adenosine triphosphate (ATP), a versatile source of chemical energy. This process must have been crucial to the early microorganisms in order to produce chemical energy in an anoxic environment. ATP is vital for metabolic processes, such as, protein synthesis and transport of molecules across membranes, and acts by lowering the energy threshold for reactions to occur.

The early atmosphere of the Earth was rich in carbon dioxide and nitrogen, so it would be advantageous for early life forms to utilise these resources in order to produce the essential organic compounds, such as, simple sugars, needed for energy. The major mechanism for producing sugars from carbon dioxide is photosynthesis. Light energy from the Sun interacts with the pigment molecule chlorophyll to provide the energy required to promote the reaction of chemically stable CO_2 molecules. Carbon dioxide and nitrogen can be made more reactive by reduction, where electrons are gained from an electron donor. Originally, hydrogen sulphide may have been used as a source of electrons for photosynthesis, since this was abundant at the time due to extensive volcanic activity.

Eventually, electrons were to be obtained from water, which, although is more difficult, is ultimately more rewarding in respect to quantum yield, and oxygen is released into the atmosphere in large quantities. Initially, oxygen was probably toxic to the early cells, due to its reactive nature. Through the course of evolution however, cells developed to exploit the reactivity of oxygen and use it to source energy via aerobic respiration. The oxidation of glucose meant that more energy is released than during glycolysis alone; as the glucose is completely degraded to carbon dioxide and water in a non-concerted mechanism, producing chemical energy, ATP, in each step. As with photosynthesis, the main energy yielding process is reduction. The acidic sulphurous conditions associated with volcanic activity, for example, at hydrothermal vents, may have provided a proton gradient across the cell membranes. This is the same mechanism, albeit primitive in design, used for reduction during aerobic respiration of cells today.

In an environment rich in oxygen, the anaerobic organisms with which life had begun declined: only some remained, living in a small number of ecological niches, such as hydrothermal vents. Others may have become predators or parasites of aerobic cells, but a few, it seems, implemented a new strategy for survival—living within the bodies of aerobic cells in a symbiotic relationship. The metabolic organisation of the first eukaryotic cells is believed to be derived from the symbiotic relationships that occurred among the early prokaryotic microbes. Eukaryotic cells vary from prokaryotes, by definition, in that their DNA is contained within a nucleus. Other distinctive organelles that are possessed by eukaryotes, and not prokaryotes, are the energy generating units: the mitochondria and the chloroplasts, which almost certainly have a symbiotic origin.

Mitochondria display many similarities to free-living prokaryotic organisms: they often resemble bacteria in size and shape; they contain loose DNA; they make their own proteins and replicate by dividing into two. The role of the mitochondria in the cell is to be responsible for aerobic respiration: a much more efficient way of producing the chemical energy, ATP, than the antiquated process of glycolysis alone. It is thought that anaerobic prokaryotes incorporated aerobic bacteria into their cytoplasm and survived by living in symbiosis: the bacteria are provided with nutrients and requite by providing ATP to the host prokaryote. In this way, eukaryotes developed as the aerobic bacteria specialised towards their function, becoming mitochondria. With the occupation of energy production consigned to the mitochondria, the plasma cell membrane of eukaryotes was now free to develop new features, such as, ion transport channels and cell signalling functions.

Chloroplasts carry out photosynthesis when sunlight strikes the chlorophyll molecules bound to the layers of thylakoid membranes contained within; similar to how the process occurs on the outer membrane of cyanobacteria. Like mitochondria, chloroplasts contain DNA and replicate by dividing into two, so the evolutionary implication is that they too originated as a prokaryote sheltered in the cytoplasm of an aerobic cell, a phenomenon that is still common today; eventually evolving into eukaryotic plant cells. Over geological time, mitochondria and chloroplasts have undergone a great deal of evolutionary change; morphologically designed to carry out their specific function: energy generation. Consequently, they have become dependent on the rest of the cell: any molecules that these organelles require are usually manufactured elsewhere and imported to the mitochondria or chloroplast.

Eukaryotic cells are around a thousand times larger than prokaryotes, which cause problems for trafficking molecules, which enter via the plasma membrane, required for biosynthetic reactions taking place within the cell interior. As cells get larger, their surface area-to-volume ratio decreases (since volume is proportional to the cube of the radius, while surface area is only proportional to the square): this compromises the efficiency of facilitated diffusion or active transport across the membrane, which are essential mechanisms for supplying the organelles with the raw materials necessary for them to exercise their complex functions. To resolve this problem, cells developed an array of internal membranes: the endoplasmic reticulum are the compartments within which proteins, lipids and other biomolecules are manufactured and exported.

Internal membranes that comprise organelles take up approximately half the volume of the cytoplasm; the remainder is referred to as the cytosol. The cell nucleus is an internal phospholipid bilayer which contains pores to allow the passage of molecules to and from the cell "control centre". Internal membranes also form stacks of flattened sacks, which are folded for a large surface area, constituting the Golgi apparatus. This structure is an intermediate factory where travelling molecules can be modified and redistributed throughout the cell.

Molecules are transported around the cell by a network of vesicles which are also formed from a membranous phospholipid bilayer, during the processes of endocytosis and exocytosis. During endocytosis, portions of the external membrane invaginate and pinch-off to form membrane bound vesicles, which move off containing substances from the external medium and molecules adsorbed onto the cell surface membrane. Even large particles, such as foreign cells, can be engulfed in this way (phagocytosis). Exocytosis is the reverse, whereby membrane-bound vesicles from inner membranes fuse with the surface membrane and release their contents into the external medium.

In eukaryotes, the DNA is contained within the nucleus; discretely segregated from the rest of the cell; isolated within the enclosed porous membrane. Over time, eukaryotes have accumulated large amounts of DNA, allowing for the observed versatility of the cells they manifest. The large strands of DNA risk breakage or entanglement, so histone proteins, unique to eukaryotes, developed to wrap the DNA up into manageable chromosomes and thereby avoid these problems. The importance of histones is reflected by their remarkable preservation throughout evolution: several of the histones in a pea plant are almost exactly the same as those in a cow. By compacting the DNA into chromosomes, bound with histone proteins, large quantities of DNA can be stored, providing eukaryotes with the potential for great diversity and complexity.

Of all the single-celled organisms, Protozoa are the most complex eukaryotes. They are evolutionarily diverse and exhibit a variety of morphologies; their mineralised ectoplasms, or tests, are preserved in the geological record and demonstrate numerous elaborate forms, developed within the confines of a single cell. Protozoa may be autotrophic, producing their own food through photosynthesis, or heterotrophic, feeding on other cells. This predatory behaviour in the early cells may have made the capture of bacteria and domestication of the mitochondria and chloroplasts possible. They have a variety of anatomical adaptations: flagella for movement, complex sensory structures, "mouth parts", stinging bristles and muscle-like contractile bundles.

The unicellular eukaryotes are grouped into the kingdom Protista, which include autotrophs (diatoms and coccolithophores) and heterotrophs (foraminifera and radiolarians). Apertures in the tests of foraminifera permit the extrusion of retractile pseudopodia, used in locomotion and catching prey. Radiolarians are planktonic, living in the photic zone of the marine environment and have symbiotic algae in their silica tests. The advanced adaptations of protists enable them to dominate this habitat.

The advancement of protists may be credited to the fact that they, unlike prokaryotes, have a nucleus. The containment of DNA within the nucleus meant that the crucial hereditary material is segregated from the rest of the cell and protected from damage and corruption. This means that large quantities, with increasingly complicated sequences, could have cellular resources allocated to it directly, enhancing the key role of DNA: protein synthesis. New, more complex, cellular bodies could be built with new features, such as cell receptors and signalling apparatus. This enabled cells to socialise and cooperate, leading to the next crucial step in the history of life on Earth: multicellular organisms.

From Single Cells to Multicellular Organisms

Single-celled organisms, such as bacteria and protozoa, have the capacity to synthesise all the substances they require from a few nutrients, so they can adapt to many environments. However, by collaboration and division of labour, it becomes possible to exploit a wider range of resources and this is the selective advantage of multicellularity. For example the root cells of a plant take in water and nutrients, while the leaves capture radiant sunlight. This enigma, where organisms use specialised tissue to collect nutrients more effectively, changed the environment, bringing a whole new dimension to ecological niches and provided an opportunity for further evolutionary change. By using specific tissues designed to perform a particular function, primary producers could collect and utilise nutrients more efficiently than in the single-cell case and consumer behaviour ensued after a time to exploit the nutrients collected by producer organisms; a new paradigm was created, where energy is transferred between organisms through food chains.

Multicellular organisms may have been derived from the association of unicellular organisms in colonies, forming clustered aggregates of protozoa that remained together after cell division. One line led to the sponges, which retained this aggregate format; the other to the metazoans, in which different parts of the body became specialised into tissues to perform a particular function, such as muscle, gut etc. Adaptations of this kind enabled organisms to move and search for nutrients, rather than being dependent on the concentration of foodstuffs at one locality; digestion in a gut improved the uptake and assimilation of nutrients, and so organisms could maximise the potential offered from the available resources and proliferate.

With the development of multicellular organisms, by the late Proterozoic, the diversification of life led to the development of the five major groups, or kingdoms, of life that persist today. These include: a) Monera (all prokaryotes), b) Protoctista (protists, nucleated algae), c) Fungi (saprophytes), d) Plantae (eukaryotic plants) and e) Animalia (sponges, multicellular heterotrophs). Organisation of cells into tissues depends on cohesion between the cells: in plants, cytoplasmic bridges remain after cell division, called plasmodesmata, and animal cells are bound by a network of extracellular organic molecules, or adhesion between plasma membranes, to form sheets of cells, called epithelium. The purpose of epithelial sheets is to enclose a sheltered internal environment. Liberty from external influences means that cells in a tissue can be kept in a consistent environment, regulated by the feedback mechanisms associated with homeostasis, so that the conditions that the cells of a particular tissue are exposed to are optimal for the tissue to perform its designated function. An early example of this biological format is exhibited by the phylum Cnidaria, which includes sea anemones, jellyfish and corals. Their bodies are composed of two epithelia; the simplest metazoan grade, termed diploblastic, where there is an outer ectoderm and an inner endoderm, which encloses a cavity that serves as a gut. The area between the two layers of cells accommodates nerve cells, which enable coordinated movement to occur.

Through the diversification of cells into specialised tissues, more complex life forms arose. Early sponges, with their uniform tissues, were the most simplistic organisms consisting of multiple cells to have existed. However, this was a profound revolution for life on Earth: multicellularity provided a new paradigm for evolution. The first sponges provided the fundamental format on which later metazoans would be created, following the evolution of cellular specialisation. Cells became organised into epithelial sheets, which enclose an internal environment regulated by homeostasis. Free from external influence, cells in the body could specialise further to a particular function, improving the efficiency of metabolism. Guts developed for digestion, muscle for movement; eventually the first complex life forms appeared.

First Complex Life
First Animal Life

Around 700 million years ago, severe global cooling caused a vast glacial event to occur, which lasted about 100 million years, during which ice sheets extended close to the equator. This period has been termed "snowball Earth". These harsh conditions prohibited advances in multicellular evolution; only unicellular microorganisms persisted. Their ability to modify promptly over generations meant that they could adapt quick enough to survive in an environment rapidly becoming more and more inhospitable. It wasn't until after the glaciation that complex life arose. It is likely that a global surge in volcanic activity ended the deep-freeze; large volumes of carbon dioxide that escaped from volcanoes into the atmosphere led to a greenhouse effect, raising the surface temperature and causing the ice to melt. Now, conditions on Earth were comfortable enough for the first animals to evolve.

Early animals were strange immobile creatures, probably white in colour, living in the deep sea. Sponges and Cnidarians were among these, but there were also a large number of peculiar creatures with no modern descendants: evolutionary experiments. During this premature stage of metazoan life, evolutionary trial and error displayed a variety of perplexing life forms, which were eventually removed by natural selection. One strange feature was that these early animals lacked bilateral symmetry as seen in most animals today: Cnidarians show radial symmetry and Sponges are amorphous. Many organisms had branching morphologies, resembling fractal patterns. These animals were common and dominated the ancient ocean floor at this time, which is preserved at Mistaken Point in Newfoundland, Canada.

Charnia is an example of one of the fractal creatures found at Mistaken Point, dating back from the Ediacaran Era, 570 Ma. The fractal pattern provided a large surface area for the absorption of nutrients and gases. These fractal fronds are similar in appearance to sea pens; modern Cnidarians. However, *Charnia* is unrelated; convergent evolution may have produced this shape independently to solve a similar problem regarding survival in a particular ecological niche, namely, efficient uptake of sustenance while being mounted statically on the sea floor. These fractal fronds may have replicated by vegetative reproduction, where parts of the frond detach and take hold elsewhere as a discrete individual. They had a wide geographical distribution and existed in large numbers, which would have taken a long time to achieve by this method of reproduction. Perhaps sexual reproduction is a better explanation: in an analogous method to corals, maybe factors such as temperature and length of day regulated the organisms and cue the timely simultaneous release of gametes when conditions were right. However, there is little evidence for this theory and it is held with much scepticism. The Ediacaran Era was probably long enough to achieve large geographical distribution, especially since the apparent success of *Charnia* from its dominant presence at Mistaken Point, meant that is was a thriving genus among the Ediacaran faunas.

The early animals, such as *Charnia*, had one feature in common: they were all anchored to the sea floor; hence lacked locomotion. Other examples of post snowball Earth, Precambrian metazoan life can be found at the Ediacara Hills in the Flinders Ranges, South Australia. This is the namesake of the geological time zone when first complex life appeared most noticeably in the geological record, 640 Ma. Preserved in sandstones deposited in shallow marine conditions, which were created as continental shelf flooded following the end of the Varangian glaciation, are a number of organisms, which presumably were capable of movement. These include jellyfish (medusoids), soft corals (pennatulacea) and segmented worms (annelids), along with many peculiar organisms of unknown phylogenic affinities, perhaps "failed evolutionary experiments".

Some palaeontologists reject the view that any of the Ediacaran fauna are ancestors of modern metazoan phyla. The organisms are classified as Vendozoa, having homogenous basic uniformity: they are thin and flattened, rounded or leaf-shaped and possess a rigid or quilted upper surface. One such example is the disc-shaped *Dickinsonia*, which had a soft-segmented body, similar to annelids. These animals varied in shape and size (1 cm – 1 m in diameter) and the segments varied as well; some showing near bilateral symmetry and others with radial tendencies. Since these organisms apparently lacked a mouth or gut, it is thought that they fed by absorbing nutrients across their underbellies. Perhaps they released digestive enzymes, as fungi do today, which breakdown foodstuffs into manageable particles. Their feeding habits are inferred from fossil evidence; unlike many other fossils from this period, *Dickinsonia* are never found overlapping or interacting in any other way, despite being abundant at the time. Where *Dickinsonia* are found in close proximity, their shapes are deformed and contorted to avoid contact in order to prevent the release of digestive enzymes onto another member of the community. "Ghost fossils" of *Dickinsonia* have been found, which may have been produced as shadows of the organism; faint imprints that may have occurred as the organism grazed on the bacterial mats on the sea floor and moved off to a new area when the food resource was exhausted.

Towards the end of the Ediacaran Era, 560 million years ago, many of the Vendozoa began to die out, including the fractal frond, *Charnia*, and the disc-shaped *Dickinsonia*. This coincided with the first appearance of more complex organisms, such as the worm-like animal *Spriggina*. *Charnia* probably died out because of its simplicity. Although the simple fractal pattern was initially successful for the establishment of early animal life, it wold have required very few genetic commands to produce; severely reducing the versatility of its DNA, therefore limiting *Charnia's* evolutionary potential because the genus could not generate novel adaptations in response to a changing environment towards the end of the Ediacaran. Consequently, *Charnia* would be easily out-competed by first complex life, which was better adapted to tackle the challenges of survival.

Dickinsonia was also unable to adapt to the new ecosystem and was replaced by more mobile organisms, such as *Spriggina*. *Spriggina* is one of the first organisms with clear bilateral symmetry: the construct of all modern complex animals.

Moreover, *Spriggina* also appears to have a head; the implication being that it had a relatively advanced sensory system for the time with which it could detect food. It is therefore likely that *Spriggina* had a mechanism to enable it to move towards the food; a direct movement in response to a stimulus, known as taxis. This was a major step for first animal life. This is in contrast to the random movement of *Dickinsonia* until it moves into more favourable conditions, known as kinesis. Taxis was a much more advantageous habit to adopt because foodstuffs could be acquired more readily, with less energy expenditure involved in random searching in kinesis. As a result the more complex organisms gathered the resources before the primitive kinds, which, without nourishment, died and ultimately became extinct.

Spriggina is thought to be an ancient ancestor of the worms, collectively known as the annelids, having a segmented body, a head, and lacking appendages. Annelids exhibit the triploblastic grade of metazoan life, where a layer of rigid tissue, called the mesoderm, separates the outer skin and the gut, enclosing a fluid-filled cavity called the coelom. Muscle contractions can push the incompressible fluid forward and permit locomotion. Perhaps this is the mechanism by which *Spriggina* moved along the sea floor and burrowed into sediment in search of food.

Post snowball Earth organisms evolved and diversified rapidly. This is probably due to the development of sexual reproduction. The mixing of genetic material vastly increased variation, thereby increasing the rate of evolution. The earliest evidence of sexual reproduction comes from *Funisia*. These columnar, worm-like organisms anchored to the sea floor lived in colonies. All of the members of a given colony were the same size, which suggests that they were different generations, which were produced at the same time and grew together as one age group. Such colonies may have been established in an analogous way to corals today, with the simultaneous release of gametes, where timing is dictated by certain environmental conditions that are optimal for breeding. Gametes belonging to two different individuals unite to create a unique offspring: this was a profound episode in the history of life on Earth because it unlocked the potential for much improved evolutionary prowess among metazoan life.

Organisms that reproduce asexually create an identical clone of the parent as a result of the direct replication of that individual's DNA. A chance mutation that yielded favourable qualities in the offspring may only be passed onto the next generation if the individual could out-divide its competitors and, so, depended on whether or not the offspring had other successful characteristics. As a result, a potentially successful mutation would only be incorporated into any given population provided that population had other favourable traits. Sexual reproduction, by contrast, involved the genes of two parents being shuffled and arranged in different combinations in each of the offspring, so, for example, a mutated bivalve would have the mutation being tested in different combinations in each of the 100,000 eggs it produced. Natural selection then operates on 100,000 prototypes, instead of only one. Therefore, in favourable circumstances, sexual reproduction can significantly accelerate evolution. Furthermore, there is an intrinsic conservatism in sexual reproduction. Extreme mutations are diluted by recombination with normal genes and remain as recessives that may reappear at unpredictable times, perhaps, by chance, when the extreme mutation is favourable for the current environmental conditions, giving that generation the selective advantage.

However, since the occurrence of sexual reproduction among certain Ediacaran faunas, such as *Funisia*, evolution would have to address the limitations of sexual reproduction. Any individual only passes half its DNA to any one offspring, so must invest twice the effort to pass off all its genes. Also, because eukaryotes have such complex genomes, two individuals must have a similar genetic makeup to successfully shuffle and recombine their DNA into a viable offspring, themselves capable of reproduction; in other words, be the same species. This necessity requires organisms to have means of identifying members of the same species in order to select an appropriate mate.

With the recession of the global winter, multicellular organisms made the most of the opportunity to proliferate. Many unique, but ultimately unsuccessful, organisms arose through evolutionary trial and error. Those creatures that were able to unite and mix there DNA produced generations that constituted a dynamic reservoir of genes, where selected combinations that yielded favourable characteristics could be produced to ensure success in a rapidly changing environment. In due course, these were the organisms that were successful, outcompeted their asexual rivals, driving them to extinction and persisted into the future. With this new mechanism to produce the next generation, by the start of the Palaeozoic, animals evolved into a rich diversity of forms.

Cambrian Explosion

The rapid evolutionary changes undergone by life on Earth during the Cambrian 505 Ma are remarkably well documented in the Burgess Shale, Lagerstätten of British Columbia, Canada. The exceptional preservation here shows extraordinary detail of even soft-bodied organisms that lived on the sea floor at that time. This has allowed palaeontologists to ascertain a near complete representation of the remarkable diversity and complexity of life that resulted from the fast-pace evolution during the explosion of life in the early Cambrian, attributable in part to the development of sexual reproduction in the Precambrian Era. Termed the 'Cambrian Explosion', the influx of new creatures included some of the modern day phyla, for example the arthropods, became established, but evolutionary experimentation also continued, producing numerous intriguing creatures, which left no descendants.

Three examples of the bizarre Burgess Shale creatures, which apparently left no descendants, are *Opabinia*, *Wiwaxia* and *Hallucigenia*. *Opabinia* had five mushroom-shaped eyes on its head along with a flexible proboscis with which it probably grabbed food from the sea floor. A broad tail and flaps down its side are indicative of its mode of life: computer reconstructions have showed that Opabinia could have wafted these flaps to generate propulsion, enabling it to swim. *Wiwaxia* and *Hallucigenia* are more typical of the Burgess Shale animals, being clad in exuberant armour. *Wiwaxia* was covered in plates and spines, resembling a 5 cm long living stronghold. The worm-like creature *Hallucigenia* had rows of spines along its arched back, which terminated at either end indistinguishably with a head and a tail. Producing armour is very expensive, requiring a lot of resources, so more nourishment must be ingested. The need for armour would indicate an increase in selective pressure on grazing animals that fed on the bacterial mats on the sea floor from predation, which is likely to have occurred as, by this time, most animals were actively collecting food and predation was the natural extension to this behaviour because it meant that predatory organisms could obtain the nourishment they required directly from another animal, rather than by competing for grazing.

The arthropods, which include crustaceans, insects and arachnids, became established in the Cambrian; exploiting the paradigm in which organisms developed armour to oppose predation. Among the first of these were the trilobites, which proliferated in the early Cambrian; successfully exploiting a variety of ecological niches; an accomplishment reflected by the diversity of their morphologies, from which the mode of life of the organism can be inferred.

The basic morphology of the trilobites, as the name suggests, is a body divided into three parts: the cephalon, which is the head shield that has further segmentation for ecdysis (moulting); the thorax and a pygidium, or tail. The exoskeleton composed of the protein chitin, which may have been additionally hardened (sclerotized) with calcium carbonate, breaks along suture lines during ecdysis, which appear on either side of the cephalon. Trilobites possess compound eyes, which consist of many calcite lenses, either side of the glabella, an inflated lump at the centre of the cephalon under which the stomach is situated. Trilobites also had antennae to further their sensory perception. Along with antennae, trilobites had other appendages attached to the cephalon that were presumably for feeding purposes. They didn't, however, have well-developed mouthparts, so probably sifted in sediment or ate small soft-bodied creatures. The thorax consisted of a variable number of segments, or pleurons, depending on the genera. Exercising the legs would have generated localised turbulence in the water, making gas exchange across the gill surface more efficient. The pygidium is made of fused segments and may have provided the trilobite with stability. Moreover, a trilobite could enrol for protection, in a similar way to woodlice, when threatened. The contacting cephalon and pygidium may have acted as a shield, covering the soft underparts.

Trilobites produced eggs from which larvae hatched in the protaspid stage, where the organism is a single segment. As they grow through the meraspid stage, thoracic segments are added; one per meraspid moult, until the holaspid stage, where the trilobite has the full complement of thoracic segments. From then on, the organism would only increase in size following ecdysis. Trilobite species varied in size from 5 mm to over 70 cm; testament to their diversity and the variety of modes of life they occupied. Each trilobite was exquisitely, morphologically adapted for their mode of life within the ecological niche.

Position

Movement

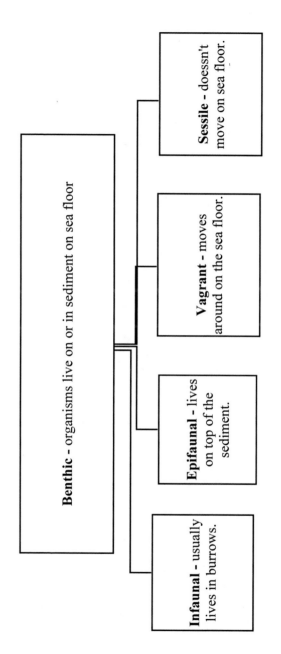

Benthic - organisms live on or in sediment on sea floor

Infaunal - usually lives in burrows.

Epifaunal - lives on top of the sediment.

Vagrant - moves around on the sea floor.

Sessile - doessn't move on sea floor.

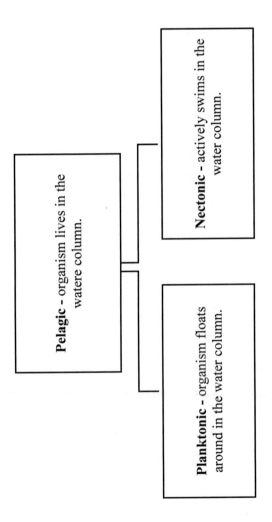

Figure 4: Modes of life of marine organisms, such as trilobites.

To illustrate the diversity of the trilobites and the variety of modes of life different species were adapted for, it is necessary to provide some examples of trilobites. *Calymene* was benthic: these trilobites dwelled at the bottom of the sea, actively hunting or scavenging. They were large and not particularly streamlined because they didn't need to swim. Their compound eyes were set high on the cheek plate, providing them with 360° vision; helpful for locating food on the sea floor. It had many pleura for leg attachment to facilitate traction on the sea floor and improve flexibility, so that it could readily enrol for protection. Some trilobites were planktonic, such as *Agnostus*, and were typically small and light, enabling them to float freely. A large glabella filled with fat or gas could have been a floatation device. They were often without eyes; as these were unnecessary for filter feeders. Other trilobites were nektonic, such as *Deiphon*. The trilobites were thought to be active swimmers because they too had inflated glabella, but the pleura were numerous and separated to allow room for many legs to oscillate; acting as paddles to propel the organism. They also had eyes: typical of an organism that locates and pursues food. *Trinucleus* was infaunal. It had no eyes, but did have sensory pits on it large cephalon, which spread its weight across the sea floor and may have acted like a shovel for burrowing.

Trilobites are an excellent example of the diversity of life within the different phyla during the Cambrian. What could have caused such exquisite diversity that stems from rapid evolution? Two ideas are the theory of punctuated equilibria and the introduction of predation to ecosystems. For the past couple of million years, life had been in a period of stasis throughout the duration of the Ediacaran. Generally, animals resembled either fractal fronds or soft-bodied discs; all of which lacked bilateral symmetry. The early Vendozoa were genetically simplistic, so evolution was slow because the potential for variation in the animals' DNA was limited; consequently changes did not visibly occur. However, a small number of animals, such as *Spriggina*, and other annelids did begin to emerge from the slow-paced evolution and were favoured by natural selection. Their ability to actively collect food, then assimilate the nutrients more efficiently in a gut, meant that they were easily capable of out-competing the formerly established communities of animals, including *Charnia* and *Dickinsonia* which, as a result, became extinct. With the removal of the Vendozoa, vacancies were left; numerous ecological niches provided the bilateral organisms with abundant opportunities to branch out and diversify, adapting to each specific ecological niche. This is probably how the Cambrian explosion began: the period through which the Vendozoa died out and the rapid evolutionary changes that followed were short in comparison to the length of time spent in stasis. However, this doesn't explain the sudden development of skeletisation.

Prior to the Cambrian explosion, no Ediacaran animals had true mineralised hard parts; only tough leathery skins at most. The abrupt appearance of skeletons in the fossil record coincides with the onset of predation. Many molluscs, arthropods, brachiopods and bivalves developed strategies of biomineralisation to build exoskeletons or shells; the armour offered protection against predators. Skeletons weren't a chance development because their dispersity among many of the Cambrian phyla is too extensive to be an incongruence of probability, nor were they triggered by chemical changes in the oceans, such as an increase in phosphate concentration, because different phyla adopt different materials to build their skeletons. Molluscs, brachiopods and bivalves typically made external shells out of calcium carbonate, albeit with different minerals; calcite and aragonite, which differ in crystal structure. Some brachiopods used calcium phosphate and arthropods use chitin impregnated with calcium phosphate.

Skeletons also support soft tissue, enabling organisms to grow larger, which incidentally also helps to oppose predation. However, it could be the case that the development of the skeleton to facilitate muscle attachment maybe the primary incentive for organisms to invest in such an accessory because of the advantages of locomotion and other benefits of having a skeleton. A box-like skeleton, the like of which brachiopods and bivalves adopted; being encased in a protective shell also helps the function of internal organs. The enclosed environment within is adapted for the sessile lifestyle, as filtering seawater is aided by the ability to generate a consistent current inside the shell and prevent clogging with silt. The belief that skeletons may have evolved initially to cooperate with the function of important organs, rather than for defence, gains validity from the fact that the early trilobites had small tails, which had limited use for protection, whereas later trilobites had larger tails that covered the underbelly when enrolled. It can be readily acknowledged that the development of skeletons enabled organisms to grow larger, but why did they? Purely speculative suggestions attribute this to global climate variations; perhaps changes in oceanic upwelling altered oxygen levels at the sea floor to a critical level where organisms were able to have skeletons, but still respire effectively given that their bodies were covered in an impermeable mineral coating, so these could no longer serve as an interface for gaseous exchange.

The shifting paradigm that propagated the accelerated evolution in the middle Cambrian generated a great amount of diversity as organisms adapted to discrete parts of the new environment. Some animals made the transition from actively collecting food to actively hunting prey; they became the first predators. This presented a new perspective to Cambrian life. Predators play a key role in the diversity of an ecosystem: predation prevents one species of prey from becoming dominant, thereby maintaining diversity in the ecosystem. It is reasonable to assume that the rise of predators may have triggered the incredible diversification that lead to the Cambrian explosion. If any one prey species became dominant, speciation occurs among predators, which evolve different traits to the ancestral population in order to hunt the abundant prey species more effectively. In response, the prey evolved new characteristics to deal with the increased pressure from predation and an evolutionary arms race is instigated from the predator-prey competition, accelerating evolution.

The elaborate armour displayed by Cambrian fauna, such as *Wiwaxia*, *Hallucigenia* and the trilobites would indicate the presence of large predators. Indeed, predation may have been a cause of the diversification of the trilobites. One such predatory animal was *Anomalocaris*; measuring up to 70 cm, it was a behemoth of its time and it is believed to be the first "big predator". Prey animals would have required extravagant defence mechanisms to repel an attack from such a powerful predator. From examination of *Anomalocaris'* anatomy, it is obviously a predator: it had two eyes on the top of its head for locating prey; two grasping appendages covered with barbs for catching prey and a spherical mouth structure that contained plates to serve as teeth for eating prey animals. Evidence from the fossilised armour of trilobites show detail of damage comparable with the appendages and mouthparts of *Anomalocaris*. Perhaps *Anomalocaris* hunted trilobites, attacking with lightning speed as mantis shrimps do today; breaking the armour of the trilobite, thereby accessing the soft body beneath.

The Cambrian explosion presents an intriguing paradox: the evolutionary radiation in the early Cambrian rapidly produced an incredible diversity of animals; propagating many new families to form discrete clades, but also an exceptional array of morphologies and body plans, that is, a great disparity. In the 500 million years since the middle Cambrian, evolution had implemented very few changes to fundamental body plans and had done so comparatively slowly. Some palaeontologists believe that the Cambrian radiation was a unique episode in the history of life on Earth and that natural selection has only tinkered with the body plans evolved in the Cambrian explosion. However, the history of life on Earth demonstrates a punctuated equilibrium: other palaeontologists argue that the radiation had to be remarkable: these were the first complex life and dispersed into hitherto unoccupied ecological niches. Among these habits was predation; a second wave of rapid evolutionary change was instigated by those animals that sought to hunt others and propagated an evolutionary arms race; a trend that would continue as predators and prey competed for survival.

The occurrence of hard parts defines the beginning of the Palaeozoic Era because the geological record suddenly becomes more complete, due to hard parts being more resistant to destruction. Indeed, for a long while, it was thought that no life besides bacteria existed prior to this era. By far the most dominant fossils are the trilobites. This is due to their evolutionary success, but also because, similar to modern crustaceans, they moulted their exoskeletons as they grew, so an adult trilobite may contribute over twenty skeletons. The diversity of their morphologies throughout the Palaeozoic Era makes them excellent zone fossils for biostratigraphy, the relative dating of geological formations. Largely, invertebrates were the dominant product of the Cambrian explosion, but a minor group of animals also evolved that were the ancestors of the major group of animals that had their skeleton on the inside of the body: the vertebrates.

Origin of the Vertebrates

One remarkable fossil found in the Burgess Shale is that of *Pikaia.* This animal closely resembles two modern day groups: urochordates and cephalochordates. Urochordates, often called sea squirts, are mobile as juveniles, but anchor themselves to the sea floor as adults, filter feeding on passing nutrients in the moving water. Cephalochordates are more complex and spend their lives swimming in search of nutrients. Cephalochordates and urochordates both have nerve chords, running laterally down the body, to which muscles are attached. Furthermore, cephalochordates have a tiny "brain"; the bundle of neurons coordinate movement by electrical signals. Fossils of *Pikaia* show a longitudinal bar that extends the length of the organism, which could be a notochord: a flexible, tough rod that runs along the length of the animal, made from a collagen sheath enclosing a fluid-filled cavity that is involved in nerve networks. The phylum Chordata is characterised by having a notochord and includes the vertebrates. In essence, *Pikaia* represents the first precursor to vertebrate life on Earth.

Pikaia has a familiar body shape that is divided into muscular segments, similar to modern day chordates, called myotomes. It is speculated that *Pikaia* had a brain akin to that of the cephalochordates, which enabled it to respond to predators in the Cambrian. Fossil evidence suggests that it sought safety among other grazers, such as trilobites. This strategy was successful if *Pikaia's* descendent were indeed the vertebrates; as this group became a dominant feature in the early Palaeozoic seas.

In recent years, a group of relatively common microfossils, the conodonts, have been suspected to have chordate affinities. Conodonts are small tooth-like structures, termed elements, composed of apatite: a form of calcium phosphate that is the mineral content of bone. They were common throughout the Palaeozoic and were found in a variety of lithology, suggesting a nektonic mode of life. In 1983, the first complete fossil conodont animal was discovered close to Edenborough, Scotland. It was a worm-like organism, with conodont apparatus (an interlocking set of conodont elements) in its head region, which were presumably feeding structures. The conodont animal was interpreted as been a chordate because it had traces of v-shaped muscle blocks (myotomes) in its trunk and post-anal tail and there was evidence of a possible notochord.

Some members of the phylum Echinodermata have certain features analogous to the chordates. The evolutionary implications of this could be that the lineage of ancestral chordates may have biological affinities to the echinoderms. This can be demonstrated by looking at a group of echinoderms called the Homalozoa, which are also known as calcichordates. Although they are generally considered to be echinoderms, Homalozoa lack the characteristic pentameral symmetry and some palaeontologists believe that they are in fact calcite-plated chordates and indeed ancestral to the vertebrates.

The features of calcichordates that led scientists to consider them as chordates are amply demonstrated by *Cothurnocytis*; a calcichordate found in the Ordovician. *Cothurnocytis* had gill slits for respiration and a tail, called the autacophore, which it perhaps used for balance and locomotion, and, critically, it may have extended along the body as a notochord. *Cothurnocytis* is thought to have lived flat on the muddy sea floor, supported by marginal spines, with the gill slits uppermost, filter feeding on edible particles.

Fossil evidence of the early chordates can be separated into three categories: cephalochordates, conodonts and calcichordates. In the early cephalochordates, a swollen bulge at the end of the notochord acted as a primitive brain; a location for the coordination of electrical signals from sensory systems, enabling a response to be carried out. These creatures eventually evolved into the craniates: a major group that makes up nearly all vertebrate life; defined as having a definite head. A true head possesses sensory organs: nose, eyes, ears, which are evolutionary advantages. With the necessary nervous connections, the cranial nerves, such as the optic, olfactory and auditory nerves make up a brain. The first craniates to evolve were the fish.

The Appearance of Fish
First Jawless Fish

In fossil records and modern fauna, there are more species of fish than all the other vertebrates put together. Fish are characterised by the possession of gills, a vertebral column and a single-loop circulatory system. They are covered in scales; have fins for balance and propulsion and more evolved forms have a swim bladder to regulate buoyancy and a sensory system: the lateral line, which runs along the body.

The first fish appeared during the Ordovician and resembled the hagfish and lampreys of today. These were the agnathans, which literally means "no jaw", as these fish lacked jaws; the course of evolution had not yet designed this feature that is common among vertebrates today. Generally, these fish had a mobile tail, a body covered in small plates and a large bony head shield made from several constituent plates. The eyes were set forward and close together and there is evidence of the lateral line sensory system that is unique to all fish, except hagfish. The lateral line is a line of open pores containing nerve endings for detecting movement in water. This was a useful adaptation for the actively swimming animals because it enabled them to sense other animals in close proximity, which may otherwise not be visible; be it to avoid a threat or pursue prey.

An example of early agnathan fish is the order Heterostraci, which are distinguished by having a single gill opening either side of the head shield in contrast to the conventional several-gill slit arrangement that was common at that time. Heterostracans were heavily armoured and lacked fins, so locomotion would have been clumsy. It has been suggested that they may have lived a benthic lifestyle, partially buried in the mud, where they would capture nearby prey by a suction mechanism of the mouth. The eyes were set forward to peer out of the mud and many species had reduced eyes because vision wouldn't be necessary in the muddy sea bottom. Some species, like *Eglonaspis*, had a long snout at the front of the head with a mouth opening at the end, which may have acted as a snorkel for burrowing or to scoop up prey from the muddy sea floor.

As with animals evolved previously, heavy armour was an evolutionary response to predation. The first fish of the Ordovician, and into the Silurian, were hunted by giant sea scorpions, called Eurypterids. These grew to enormous sizes; the largest of all the arthropods, with some species growing to three meters in length. It is possible that sea scorpions grew to such incredible sizes because of the escalating effect of the evolutionary arms race between predators and prey that started in the Cambrian. Large size would enable Eurypterids to capture large fish and thereby establish their role as apex predators.

Eurypterids are members of the phylum Chelicerata. Like other members of the phylum, such as horseshoe crabs and spiders, Eurypterids are characterised by having a body of two parts: the prosoma, where the head and thorax are fused together, and a segmented opisthoma, or abdomen, which terminates into a telson of various morphologies between species; examples include paddles, spikes and pincers. Sea scorpions, such as *Pterygotus*, would steadily stalk prey, including early fish, trilobites and even other sea scorpions; walking along the sea floor and, with a thrashing a the telson to attain a burst of speed, capture the prey item, gripping it with serrated pincers called chelicerae. *Pterygotus* would then persist in tearing the prey item apart with its chelicerae into pieces small enough to fit into its comparatively small mouth. Eurypterids had six pairs of appendages attached to the prosoma: two were the food-handling chelicerae, four stout walking legs close to which the gills were situated and, finally, two paddles used by the scorpions for swimming in a breast-stroke manner; a more efficient means of locomotion through the liquid medium than walking.

Despite their name, sea scorpions often lived in fresh water. More significant was their ability to survive, for short periods, outside of the water and, for the first time in the history of life on Earth, set foot on land. In some species the gills evolved as specialised vascular tracts of the underbody protected by the appendages and would have, perhaps, enabled the gills to retain moisture, enabling eurypterids to walk on land for short periods. These animals were true evolutionary pioneers and set the way for the arthropods to conquer land and create a new evolutionary paradigm.

Progressing from the early vertebrates, the first fish evolved as comparatively efficient swimmers. Initially, the notochord in the early vertebrates such as *Pikaia*, and later in the first fish, acted as a stiff rod at the centre of the body to hold energy from the body muscles and then release it in a rhythmic sinuous motion to generate propulsion and enable efficient active swimming. The efficiency of swimming in this way increases with body length, so the first fish became more slender to reduce hydrodynamic resistance and to produce a stronger swimming motion. Beyond a certain size, though, swimming requires more rigidity than can be provided by a notochord; this instigated the evolution of the backbone. However, many of the first fish to appear in the Ordovician, in fact, used mineralised outer body plates rather than an internal skeleton to support the body. This is most prominent on the head shields, as with *Eglonaspis*, and may be a vestige of the armour response in the continual evolutionary arms race between predators and prey that became prominent during the Cambrian.

Vertebrates used calcium phosphate as their hard-part mineral rather than calcium carbonate, calcite. Fish, humans and all vertebrate are alike in that they are capable of producing powerful bursts of energy, but this involves breaking down sugars in muscle cells faster than the oxygen that is requires to do so can be absorbed; consequently, there is an oxygen debt. Anaerobic glycolysis produces lactic acid; a build-up of which causes fatigue and the oxygen debt must be repaid during rest, when the lactic acid is broken down. Consequently, the acidity of the blood varies over periods of activity and rest and calcium may be dissolved temporarily out of the bone. If a skeleton was to be made out of calcium carbonate, rather than calcium phosphate, too much of the skeleton would be dissolved. Phosphate is dense, so the heavily plated fish would most likely have swum clumsily close to the bottom of the sea.

Efficient active swimming was a great evolutionary success. It was, and is, a means of locomotion with conservative energy expenditure, so vertebrates could travel further in search of food and also generate speed readily to avoid predators. Predators, therefore, would have to adapt to capture these prey types. Most notably, the Eurypterids evolved to be the dominant apex predators; using huge size and ferocity to generate power to overcome the first fish. Active swimming in open water was essentially an entirely new ecological niche in the Ordovician, so abundant opportunities were available for the first fish, which consequently were able to flourish, moving into the Silurian and particularly the Devonian, with the evolution of improved feeding mechanisms: jaws.

The Golden Age of Fish

The Devonian, ranging from 408–360 million years ago, was truly the golden age of fish. Throughout its duration, fish evolved and diversified a great deal, becoming the dominant animals in the water. They exhibited a lot of variation, filling a range of aquatic niches, but also set the path for the evolution of new vertebrate groups. At the beginning of the Devonian, fish undertook a major evolutionary step and became the first vertebrates with jaws. These fish were the gnathostomes. This was a very successful feature: it enhanced the predatory mode of life of fish by enabling them to catch and securely grip prey. Jaws also enabled fish to adopt new food-handling techniques, allowing the food to be manipulated, cut cleanly and grounded up; thereby improving the efficiency of digestion.

The evolution of jaws is thought to be derived from the gill arches. The gill arches are paired bones that hold the separate gill slits open. It is thought that the furthermost anterior pair became the teeth; the third pair, the upper and lower jaw, and the fourth pair thickened to become part of the supporting skull. Some palaeontologists believe that the origin of the jaw, rather than to function as a food-capturing device, was to generate counter-current flow, where the passage of water across the gills is in the opposite direction to blood flow in the capillaries in the gills, therefore creating a diffusion gradient, making gas exchange across the gills more efficient.

Evolving jaws from the gill arches was an important derived feature for the gnathostomes because it enabled them to source a much more diverse choice of food; hitherto, agnathans were confined to eating small particles, such as plankton and soft-bodied animals like worms and jellyfish. By the Devonian, fish were capable of hunting larger prey; invertebrates and other fish were a common food source and this diet would become important; being staple of modern aquatic food chains as well as for vertebrate evolution beyond the water. However, feeding may not be at the heart of jaw origins. The concept of counter-current flow may be the primary advantageous characteristic derived from having a functioning jaw, with mechanical feeding apparatus being a later modification. The gill arches, made of thin strips of cartilage or bone, support the soft gills against the current as the fish swims through the water. The more water passes the gills, the more oxygen can be absorbed and the fish can generate more metabolic energy. Most living fish use a pumping action where they increase and decrease the volume of the mouth cavity by flexing the jaws. Hence, if this mechanism was derived in the early jawed fish, then it is likely that the evolution of jaws was connected with respiration rather than feeding.

Logistically, modification of the primitive structure of the respiratory systems in fish as the primary reason for developing jaws would probably have been more prudent than to have designed jaws predestined as feeding apparatus. This reasoning stems from the issue of energy generation. By the Devonian, fish began to increase in size, so gills alone became inadequate to supply a large body with oxygen. That said, the earliest gnathostomes were mostly small fish, but the important aspect of their ecology is that these fish started to evolve towards modern habits. Having jaws that increased the effective current of water passing the gills, by implementing a pumping mechanism, enhanced the amount of oxygen that the fish could absorb into the body. Oxygen debt was no longer a limiting factor on the amount of activity a fish could perform; more metabolic energy could be generated to fuel the muscles. Consequently, these fish could swim faster and successfully pursue speedy prey, such as other fish. Naturally, the next step to improve hunting performance was to modify the jaws slightly to a system that could seize the speedy smaller fish. Eventually, the gnathostomes became the dominant group of fish.

Among the oldest known gnathostomes; the acanthodians evolved in the Silurian and became more common in the Devonian, when jawed fish really began to emerge. The acanthodians were generally small fish, mostly less than 200 mm, and their name refers to the spines along the body. For example, *Climatius*, had two dorsal fin spines. The spines were defensive structures, making the small 70 mm *Climatius* uncomfortable to swallow by a predator. These scales on acanthodians, made of bone and dentine, grew at the margins as the animal grew larger. Similarly to sharks, acanthodians had cartilaginous skeletons and, like many sharks, were predatory, having teeth and large eyes; characteristic of deep-ocean predators. Moreover, fossil evidence has been found of a small fish within the body cavity of an acanthodian.

Another primitive group of jawed fish that evolved in the Devonian were the placoderms. The class Placodermi is divided into nine orders: the largest group, which incorporates more than half the known placoderms, is the order arthrodia. While the head and front portion of the body of the placoderms was armoured similarly to the Heterostracans before them, the placoderms were different in that the presence of a neck joint enabled the head to be lifted. A gap between the trunk shield and the head shield (the nuchal gap) enables the mouth to open by means of an upward swing of the skull and a dropping of the lower jaw. This mechanism of opening the mouth is supposed to be advantageous to the placoderms because of their bottom-dwelling predatory lifestyle: this was the most effective way to capture prey along the sea floor. The arthrodires in the late Devonian achieved enormous size, such as the 10 m long *Dunkelosteus*. Beak-like teeth formed from the edge of the jawbone were perfect for crushing and puncturing, making them the dominant predators of the time and the largest vertebrates yet to evolve.

The first sharks and rays, or chondrichthyes, meaning cartilaginous fish, are generally recognised as the most primitive living gnathostomes and indeed, jawed vertebrates. Body fossils of sharks are rare because preservation of the cartilage skeleton is uncommon, but numerous fossil teeth are preserved, enabling palaeontologists to interpret the behaviour, diet and size of the sharks throughout geological time and ascertain evolutionary trends. Sharks started off as small predatory fish in the early

Devonian and increased in size and diversity as they evolved over time.

An example of one of the early sharks is *Cladoselache*, which reached lengths of up to 2 m, so was larger than many primitive forms that perhaps had closer acanthodian affinities. The tail of *Cladoselache* is externally symmetrical, but internally the notochord bends upwards into the dorsal lobe. It has two dorsal fins: one behind the head; the other halfway down the body. There are two sets of paired fins: pectoral and pelvic fins, each associated with a girdle element of the skeleton. *Cladoselache* was probably a fast swimmer, using sideways motions of the broad tail as a means of propulsion and the pectoral fins for steering and stabilisation.

The evolution of sharks continued at a fast pace after the Devonian: many groups radiated during the carboniferous, including the lineages of *Cladoselache*. Some sharks of the carboniferous had peculiar appearances, such as flat extensions of the dorsal fin, perhaps as a defensive mechanism or for mating displays. Neoselacheans, including modern sharks, evolved in the Mesozoic and diversified through the Cainozoic to include 35 families. These sharks were characterised by numerous features: jaws that open more widely than in earlier forms due to more mobility in the joints, the snout is longer than the lower jaw, so is more suited to tearing flesh from larger prey, and the notochord is encased in calcified cartilage vertebrae, whereas previous chondrichthyes lacked calcification. Among these was a relative of the great white shark, called *Charcharadon*, which is known due to its 150 mm long fossilised teeth. A graphical plot of tooth size against body length of living sharks enables an estimate from extrapolation of the body length of *Charcharadon*, which is predicted to be a colossal 13 m long, compared to the 6 m long great white. However, estimates of this kind can have large discrepancies; perhaps the prey that *Charcharadon* targeted required it to have developed proportionally larger teeth than living sharks. Nevertheless, sharks became a very successful group of fish and this is exemplified by their preservation through evolution over 320 million years, changing very little, being perfectly adapted to their natural habitat.

The Bony Fish

The early bony fish, belonging to the class osteichthyes, also arose noticeably in the Devonian and, from the beginning, separated into distinct subclass lineages: the actinopterygians (ray-finned fish) and the sarcopterygians (lobe-finned fish). As the names may suggest, these fish were distinguished by the structural morphology of their fins: ray fins are supported by bony/cartilaginous rods; lobe fins by a single basal bone to which muscles were attached that could modify the posture of the fin. Fish evolution after the Devonian followed two main paths: the chondrichthyes radiated several times and the osteichthyes became very diverse. Today, actinopterygians, or ray-finned fish, constitute about 95% of modern fish, examples including salmon, cod, herring, goldfish and tuna. Lobe-finned fish continue to exist today, but are much less common. However, the evolutionary lineages of the sarcopterygians display the crucial stages of vertebrate evolution regarding the first appearance of vertebrates on land.

An example of a ray-finned fish from the middle Devonian is *Cheirolepis*. Typically 250 mm in length, *Cheirolepis* had a slender body with a heterocercal tail; although the tail fin beneath made it almost symmetrical. It was a fast predator, presumably using its large eyes for hunting. The head skeleton was composed of several mobile units, which adjusted position to enable the fish to have a large gape and capture fish up to two-thirds its own length. The later radiation of the actinopterygians, like *Cheirolepis,* meant that they would become the dominant fish in the sea today.

Sarcopterygians, or lobefins, were a much more significant group in the Devonian, but have since become rare. They are characterised as having muscular pairs of lobe fins, with bony skeletons. The sarcopterygians radiated into four principal groups that were common in the Devonian: the osteolepiformes, porolepiformes, coelacanths and lungfish. Coelacanths were thought to have become extinct at the end of the Cretaceous until the discovery of *Latimeria* in the Indian Ocean. Lungfish exist today, but only in three genera.

A very significant stage in vertebrate evolution is the lineages of the osteolepiformes, which are thought to be ancestral to the first tetrapods, which eventually evolved lungs that were capable of supporting the body outside of water. The ability to walk on land was derived from the utilisation of the paddle-like fins to thrash around in shallow water. This was a very important innovation in the fossil record because it offered a large number of ecological niches to exploit on land. One lineage of lobe-finned fish to do this are known as the rhipidistians.

The development of air-breathing in rhipidistians and lungfish may have been derived in response to, or even as an indirect consequence of, the conditions in which these fish habituated, living in oxygen-poor waters, such as lagoons, along shorelines and in tropical lakes. Seasonally low oxygen concentrations in water require fish that live there to develop mechanisms to improve oxygen intake. Respiration through gills can be inefficient: fish have to actively pump water over them, which requires more oxygen to produce energy and free space in open water is necessary to swim at a fast pace to generate enough of a current. In aquatic environments where oxygen is in short supply, fish have adapted by consuming air bubbles through the mouth to serve as an oxygen store, since oxygen can be in a higher concentration in the air bubbles than dissolved in the water of lagoons. The waste carbon dioxide can then diffuse into the bubble from the gills and be ejected from the mouth. This strategy may seem far removed from true air-breathing, given that the gills dry out in air and collapse therefore don't work out of water. However, an extension of the air-bubble strategy was to evolve a mouth pouch, which served as a primitive lung. Air entered the pouch through the mouth, oxygen was passively diffused into the tissue and carbon dioxide expelled, then an air-pumping system was applied, say a muscle contraction or closing the mouth, to collapse the pouch and evacuate the stale air content. Rhipidistians had nostrils to evacuate the air, but also had gills, so they had alternative ways of breathing. As these animals became more terrestrial in habit, gills were omitted.

Land affinities in rhipidistians and any other sarcopterygians, initially, may have evolved from modification of the characteristic lobed fins rather than respiratory adaptations. Movement in the muscular fins that could change posture enabled these fish to pull their bodies across solid surfaces, such as in the shallows or onto the land's surface. These fish could then have evolved a tolerance to air secondarily by using lungs, which permitted them to spend more time on land.

One hypothesis for the modification of lobefins to limbs is that the fish could move from a dry pond during a drought to another pool. The ability to withstand exposure to the air would, no doubt, help to do this. However, research has shown that animals tend to stay in drying pools because the chance of survival is better than making an excursion across parched land to find another pool. Also, rhipidistians tended to live on shorelines rather than in inland seasonal rivers, so they didn't have to endure through long dry periods. An alternative theory is that basking became important. Rhipidistians were preadapted to air exposure when crawling over mud banks in search of food; the primitive lung system described above may have evolved as a result of these excursions. If they evolved the habit of sun bathing to warm their bodies, digestion would have been faster, so they may have grown faster, matured earlier and become reproductively more successful than their competitors that didn't bask. Initially, only the back would be exposed, extruded from the shallow water, but the effectiveness of basking would have encouraged more complete exposure. With the loss of buoyancy from water, more weight was on the body: the risk of suffocation because the thorax couldn't contract became a problem. Most of the weight would be on the pectoral fins, which became stronger to hold the body up. As the strengthened lobefins developed to become more mobile; with improved dexterity, locomotion on land became more efficient.

Another incentive to vacate onto land is reproductive security: spawning in isolated ponds, free from predators, would significantly increase the chances of survival to adulthood; as eggs and hatchlings are the most vulnerable stage of life for a fish. Also, small ponds are ideal breeding sites for small invertebrates, providing a rich food supply for young rhipidistians. Once large enough, the young could make their own way back to the adult habitat; walking maladroitly from their rock-pool nurseries to the seashore.

During the Devonian, vertebrate evolution began to excel. Prior to this period, most of the profound evolutionary changes were among invertebrate phyla, notably during the Cambrian explosion, where the first complex animals began to radiate. At this time, the vertebrates were only represented by a small number of notochord-bearing organisms, such as *Pikaia,* which were worm-like and swam by sinusoidal muscle movements, supported by the stiff longitudinal chord down the spine. The first jawless fish, the agnathans, evolved as the primitive chordates began to develop bony plates of armour, primarily to make a headshield. These first fish were restricted to sifting through sediment to find soft-bodied prey, with minimal physical exertion, that could be managed without chewing. The evolution of jaws in the gnathostomes enabled more efficient respiration, and a more active lifestyle as a result. These advanced fish could pursue prey more speedily than their rivals, but also manipulate the food in the mouth, which offered a wider range of dietary options. Ultimately, these fish became the most successful and dominated in the Devonian. The golden age of fish also saw some clades of bony fish with lobefins, the osteolepiformes, evolve to breathe air and adapt their paddle-like fins to become true limbs. These became the first amphibians and signified a new chapter of vertebrate evolution in the history of life on Earth: the conquest of land.

Life on Land
The Conquest of Land

The conquest of land was a major step in evolution and one of the most significant events in the history of life on Earth. It enabled further diversification of living beings as they adapted to new environments and occupied the rich variety of ecological niches that, hitherto, did not exist. Plants were the pioneers of the hostile frontier, making an existence on the harsh, sterile terrain, around the margins of water bodies. Communities of plants provided an environment in which arthropods could live, without having to return to the water after short exposure to air. Once arthropods became established, they provided a source of food for fish that had the necessary adaptations to move around on the shoreline. After a time, functioning ecosystems were established on the Earth's terrestrial surface.

Although some arthropods, such as the eurypterids, were accustomed to leaving the sanctuary of water for short periods, initially, it was impossible for animals of any kind to live on land. Animals did not yet have the means to extract oxygen from the dry air, ultraviolet radiation from the Sun was damaging to their cells and, originally, no biomass existed on the sterile land: animals simply had no food. In order for animals to live on land permanently, conditions had to improve.

Plants, by contrast, were a different story: they were able to utilise sunlight to synthesis their own food. Like animals, plants also experienced fierce competition; in the Palaeozoic seas fighting for access to light and nutrients. The intensity of light is significantly reduced as it enters the water because a great deal of the light is reflected off the water's surface and the intensity decreased further with depth due to scattering of the light. Sunlight can only penetrate so far, in a region known as the photic zone, which is where photosynthesising organisms reside. Any territory that a plant can claim in the shallow water, where light intensity is greatest, is extremely valuable. For the purpose of maximising photosynthesis, plants have a tendency to grow towards the sunlight; a behaviour known as phototropism, hence, occupying regions close to the water's surface, where there is intense competition as a result. The intensity of light is greatest above the water, so any plant that could extrude photosynthesising cells above the surface of the water would be at a selective advantage. Not only would this be advantageous to the plant for collecting energy and photosynthesising food, the carbon dioxide required by plants for photosynthesis is more readily collected from the air than water. The plants could therefore carryout more photosynthesis, producing more sugars, and grow to reproductive maturity faster and pass their successful genetic material to the next generation.

However, plants had to experience challenges as they made the transition from an entirely aquatic lifestyle, and had to adapt accordingly. Plants had to make changes to their physiology in order to function in the air medium and cope with the associated problems. One of the main problems faced by aquatic plants was water loss. Since this wasn't an issue for plants submerged in water, they were not adapted to be conservative with water. Pores under the leaves, called stomata, permit the exchange of gases, including water vapour. Plants had to develop the ability to close the stomata to prevent water loss during hot, dry periods. This is done by using turgor pressure in the kidney-shaped guard cells present on either side of the stomata. When water is scarce, the guard cells become flaccid due to the lack of water and close over the pore, so that further loss of water through transpiration is reduced. Further adaptation to prevent water loss due to transpiration is the waxy cuticle: an impermeable layer on the leaf's surface.

Another problem that arose as plants ceased to be completely submerged in water was the need to transport water to the emergent tissues. Parts of a plant that are exposed to the air needed water for photosynthesis and if there is no means of transporting water to those tissues, the plant risks desiccation. Early semi-aquatic plants were small enough to not need leaves, roots or elaborate water-transport systems and, perhaps, adopted techniques similar to those of modern mosses. Instead of having a system of hollow tubes, the cells of mosses are aligned in a grid-like format. When the photosynthesising tissues at the top become dehydrated, water moves by osmosis from the saturated tissues at the bottom to replace the water lost through transpiration. Although this is a simple solution to water transport, it is inefficient and restricts mosses to wet environments, and limits the size that they can grow to. Due to its simplicity though, this method could have been used to hydrate the emergent tissues of the early semi-aquatic plants, and perhaps the first true land plants also.

The biology of a typical plant could be derived from this semi-aquatic ancestry. Plant cells in the water were probably depended on the surface tissues for the sugars produced during photosynthesis, while the surface-dwelling tissues required the water and nutrients taken in by the submerged part of the plant. Over time, these circumstances probably caused the cells to specialise: cells that collected water became the earliest root cells and assembled into dendritic tissues with a large surface area, and the photosynthesising cells became the first leaf cells, which assembled into thin broad laminated tissues. Hypothetically, this could be how the first land plants developed.

The passive diffusion of water by osmosis severely limited the height to which a plant could grow, holding it at a disadvantage for the competition for radiant sunlight. A new mechanism for water transportation was required, now that plants began to completely abandon their aquatic homes. Previously, plants had no need for such a system: all the water and dissolved minerals they required were readily available in the water around them. Land plants, by contrast, had to transport water from the specialised roots to the specialised leaves, where it was needed for photosynthesis. This became an increasing distance as plants grew taller, due to the necessity of phototropism, therefore a more efficient mechanism was needed than the currently implemented passive osmosis strategy. Plants evolved a system of tubes through which water could pass, called xylem vessels.

Water is extracted from the soil by a network of roots. The outermost layer of root cells have long projections: these are the root hair cells, which increase the surface area for osmosis; water diffuses into the cell because it is a region of low water potential, due to a high concentration of sugar and ions in the sap. The water then moves towards the xylem vessels, located at the root core, by osmosis in a similar way to mosses, and the speculated method of water transport in the early plants. Water has two means of moving through the root tissue: it either travels through air spaces in the cell walls along the apoplastic pathway, or through the cytoplasm, passing through gaps in the cell wall called plasmodesmata, along the symplastic pathway. A waterproof layer, called the Kasparian strip, redirects all the water to the symplastic pathway before it moves across into the xylem. In order to enter the xylem, mineral ions must first be actively transported to lower the water potential inside, so that water can enter by osmosis. This part of the process requires energy. Because of its cohesive nature, water then moves up the xylem, due to the pull from transpiration, until it reaches the dehydrated tissues that require it for photosynthesis.

In turn, the sugars produced in the leaves by photosynthesis are transported to the cells that require the sugars to generate chemical energy, in the form of ATP, through respiration. This energy can then be used for the active transport of mineral ions and thereby continue the movement of water into the xylem. The sugars are transported down the phloem tubes, which run parallel with the xylem vessels, in a process called translocation. The columns of xylem vessels and phloem tubes are collectively called vascular bundles.

Fossils of the first land plants from the Silurian show evidence of plants having developed these elaborate networks of vascular bundles. The best example is of a specimen called *Cooksonia*. *Cooksonia* was among the first terrestrial plants: it had photosynthesising cells arranged in branches, rather than leaves, and had a rudimentary root system. *Cooksonia* was a well-developed terrestrial plant. Presumably, there were primitive plants with aquatic connections, that showed the transition of plants from water to land from before the late Silurian/early Devonian, but fossil evidence is scarce.

Cooksonia, however, lacked some of the characteristics associated with modern land plants, such as a waxy cuticle and stomata that are capable of opening and closing, and would have suffered from unnecessary water loss. To compensate, it would have been restricted to water-drenched locations, namely, tidal zones. As the early land plants colonised these locations, the crowding would have resulted in competition. Natural selection resulting from the competition would have caused plants to adapt in such a way as to alleviate the pressure from competition by migrating from the shoreline locations: pioneer species would have colonised the free space further inland. By the Devonian, plants had developed desiccation-resistant traits, such as cuticles and controlled stomata, to cope with the drier environment.

Low population densities in the more arid environments meant that plants had access to the maximum amount of sunlight, so they were able to thrive and reproduce, passing on their xerophytic adaptations, coded in their genetic material, to the next generation. This produced the next wave of xerophytic species, that were adapted to arid conditions, and eventually the ground would again become crowded. To out-compete the plants around them, some plants grew taller. Specimens of *Cooksonia* were only 1–2 cm tall, but some Devonian plants, by contrast, grew to a meter in height and had leaves on their stems to collect sunlight more efficiently, while simultaneously shading their neighbours; reducing the amount of sunlight available to them. However, the increase in height reduces the stability of a plant, which had problematic implications.

To overcome the lack of stability and the risk of toppling, some plants, the progymnosperms, evolved a hard layer of woody tissue around their stem, in order to improve upon their structural integrity. With the rigid tissue around the stem and the development of extensive root systems to serve as anchor points, progymnosperms could grow much taller. These plants were the ancestors of redwoods, pines, conifers, spruce and many other tall-tree boreal species. Through evolutionary progression, woody progymnosperms, during the mid-Devonian, specialised their leaves further by developing fern-like morphologies, which vastly increased the surface area of the leaf and therefore the amount of photosynthesis that the leaf was capable of producing. This reduced the need for plants to grow new stems, which significantly conserved energy resources.

Having leaves, wooden bark and extensive root systems and reaching heights of up to 10 m, progymnosperms were essentially the first trees. As small copses expanded through ecological succession, they amalgamated to become the first forests and other plants, which could survive in lower levels of light, became the first forest ground vegetation. In short, an entire new biome evolved composed of multiple constituent ecosystems: a whole new evolutionary paradigm.

Despite these new evolutionary characteristics, plants were still dependent on abundant moisture: specifically for the business of reproduction. This meant that plants continued to be restricted to damp conditions, such as swampy forests radiating from water sources. Over geological time, degradation of the biomass from these environments, through the process of diagenesis, produced deposits of coal. This is particularly prevalent in carboniferous formations. The reason that plants relied on these wet conditions is because they had to release their gametes into rainwater or moisture collected on the leaves, so that the sex cells could be able to travel through the water and engage in fertilisation, producing an embryonic plant.

The arid conditions of the Devonian made it difficult for plants to reproduce away from water sources, where humidity is high. Because of the unexploited space in the dry environment, plants had an opportunity to diversify; natural selection would favour species that could produce saplings without the perpetual need for humidity. Evolution's solution to this dilemma was to produce structures analogous to the first seeds. Seeds are the hard coverings that almost completely encase the embryonic plant. There function is to provide protection and store starch to provide the embryo with energy, so that it can develop roots and leaves to become self-sufficient at collecting energy and nutrients. Significantly though, the protective covering substantially reduces water loss; circumventing the problem of desiccation. With the release of sperm in the form of pollen, plant eggs enclosed in the safe capsule could be fertilised and the resulting embryo would be able to grow; free from the necessity of water.

With liberation from the need for incessant moisture, the evolution of the seed-bearing gymnosperms enabled plant successions to occur in more arid environments. Plants were able to colonise new land: different species existed in different environments, where adaptations were complementary with the ecological niche. Slowly, biomes propagated; the versatility of plants meant that they would come to dominate the landscape.

Looking at ecological successions occurring at present and applying the concept of uniformitarianism, it may be assumed that plant succession of the past occurred in a similar way as they do today. It follows that, once plant communities have climaxed and provided a sustained habitat, animals will adapt to fill the vacant ecological niches. From around 450 million years ago, the first land animals appear in the fossil record: arthropods move from the seas to utilise the resources found in the recently established plant communities. As plants did, at the beginning, arthropods would have experienced challenges in the transition to life on land.

During the Silurian, it is thought that arthropods, such as eurypterids, emerged onto land for short periods. Certain problems, though, prevented them from remaining on land for longer intervals. Initially, there were no plants on land, so no food was available for the arthropods. While some species of eurypterids had adaptations to keep their gills moist for short periods on land, arthropods would have been susceptible to sustained water loss and, consequently, would dehydrate if they remained out of water. Gills, of course, are designed to extract oxygen from water, so arthropods would have to evolve a mechanism to enable them to breathe air. Arthropods would have to overcome all of these problems in order to conquer land.

One unique fossil find is that of *Pneumodesmus newmani*; a myriapod (millipede or centipede), measuring one centimetre in length. The significance of this fossil is that it has a line of pores down its flank, which could indicate that it was an air-breather. Arthropods did indeed evolve a novel air-breathing system: holes in the body, called spiracles, that allow air to passively diffuse into the body through a network of tubes, or trachea, which branched out further into tracheoles that extend throughout the animal's body to provide a large surface area for gaseous exchange and short diffusion distances to maximise the efficiency of ventilation. Respiratory gases move through the tracheal system in two ways: along a diffusion gradient, created as oxygen is used by the cells and carbon dioxide is produced; and by ventilation, where volumes of air are moved as the arthropod contracts its muscles. Due to the reliance of this system on diffusion, gas exchange pathways must be small, which limits the size of terrestrial arthropods, including the insects of today.

Despite the colossal size of some eurypterid species, which enabled them to become apex predators in the Silurian, small size was not a handicap to the first terrestrial arthropods. A small volume and a small surface area meant that they could survive with less water and lost less water from the body through respiration and evaporation. Additionally, the exoskeleton, which was strong enough to support the body out of water, had a waterproof covering, or cuticle, to further reduce water loss. Eventually, adaptations were evolved that would reduce water loss even more, such as valves over the spiracles, so that they could be opened and closed selectively to regulate water loss.

Dating of the rock, in which the fossil of *Pneumodesmus* is preserved, found its age to be 428 million years; from the mid-Silurian. It is likely that *Pneumodesmus* was among the first animals to live a fully terrestrial life. What caused millipedes and other arthropods to adopt a terrestrial mode of life, though? Current theories are derived from the interconnection with the appearance of the first land plants. One explanation is that arthropod populations bloomed when plentiful food became available in shallow waters from the remains of rotting plant matter at the water's edge. As the population grew, competition for food increased. Individuals that could withstand feeding on the decaying plant material onshore for longer intervals would be at a selective advantage: they could feed more, reproduce more often and, therefore, pass on their genes; eventually modification with descent enabled their descendants to live a fully terrestrial life. Another explanation is that rotting plant material generated anoxic conditions in the shallow water. This would force the arthropods living there, towards the shore, where there was food available, in order to avoid asphyxiation. The selective advantage was acquired by those arthropods that could access the most food on land.

Another theory as to why arthropods found it advantageous to emerge on land stems from the pressures of predation. Firstly, land would provide a safe-haven and respite from the persistent oppression instigated by marine predators. Arthropods that could spend the greatest length of time on land reduced the chance that they may be eaten by a predator. Those individuals lived longer lives, reproduced more and dispersed their successful genes throughout subsequent generations. Secondly, since evolution is propagated through the preservation of successful genes, it is logical that arthropods would strive to protect the next generation in its premature vulnerable stages. By laying their eggs on land, which was free of egg predators, the young could have a safe nursery where they could develop and mature. Through evolutionary progression, this affinity with land may have developed into a more long-term basis. If the arthropod larvae were safe on land, perhaps it was advantageous to remain there for mutated forms, which adapted to tolerate the terrestrial environment, to complete their life cycle and produce subsequent generations of pioneers that endeavoured to utilize the abundant resources presented by the established terrestrial floras.

How arthropods came to adopt life on land is unclear, but they did indeed embark on a terrestrial mode of life, and did so very successfully. They evolved and diversified; their populations grew as they adapted to fit in the variety of ecological niches presented by the recent terrestrial vegetation. In this era of the history of life on Earth, they were the only land animals; arthropods dominated the available habitat and were free from the predators that targeted them in the marine environment. The presence of numerous arthropods on land, particularly frequenting shorelines during the Devonian, is likely to have drawn fish towards the shoreline to exploit this new food source. One particular groups of fish, the sarcopterygians, may have been able to access this food source; predating on the arthropods living in the shoreline detritus. The muscular fins positioned low on the body of the lobe-finned fish could probably enable them to move around on land, albeit laboriously, in a similar fashion to the mudskipper today. It is likely that these lobefins also had similar respiratory adaptations to mudskippers, being able to oxygenate their blood, to a small extent, via the passive diffusion of oxygen through the skin.

Evolutionary adaptations to aid gas exchange out of the water would have been derived from the hunting techniques displayed by sarcopterygians. Prey animals, such as small fish and arthropods, would have sought sanctuary in shallow water. Sarcopterygians, with their mobile fins, would have been able to drag themselves along the riverbed or seashore to access these prey animals. As arthropods began to emerge on land, living in the shoreline detritus, the sarcopterygian fish would follow; hauling themselves onto land to feed on the arthropods there. Any fish capable of staying out of the water for longer periods would be at a selective advantage because it could spend more time feeding on the abundant, virtually competition-free food source on land. It follows that fish evolved specialised organs to extract oxygen from the air. Fossils of the genus *Eusthenopteron* had nostrils at the front of the head, connected via nasal canals to the mouth, and are thought to be among the first air-breathing vertebrates.

Before making the transition to life on land and embracing an amphibious lifestyle, these fish had to overcome the challenges associated with terrestrial living. In addition to being physiologically capable of breathing air, fish had to deal with the issue of supporting a body out of water. The buoyancy of water supports a fish's body, so its weight is negligible with respect to the structural mechanics, but in order to cope with the stress of gravity, the skeleton and the internal organs had to be structurally modified. Changes to the vertebrae had to be made so that the backbone would hold the body, preventing it from sagging between the limbs and, indeed, the limbs had to develop to carry the body as well as producing a mechanism of locomotion across land.

The early evolutionary origins of limbs can be seen in the Devonian sarcopterygian fish. The osteolepiform *Eusthenopteron* had pectoral fins comprising the major bones of a tetrapod (literally meaning four feet) limb: the single upper arm bone, the humorous, the two forearm bones, the radius and ulna and the main wrist bones. The osteolepiform pelvic fin is also anatomically analogous to a limb, containing the principle bones associated with a hind limb: the thigh bone (femur), the lower leg bones (tibia and fibula) and the major ankle bones. Despite being anatomically similar, osteolepiform fins and tetrapod limbs functioned differently: *Eusthenopteron* could not walk effectively on land.

The limb bones in *Eusthenopteron's* lobe fin point backwards and movement was primarily focused on the shoulder joint. This arrangement was suited to the fish's mode of life:

swimming and mud-skipping in the shallows. The limb bones had to undergo a morphological rearrangement in order for tetrapods to walk on land effectively. Structural changes were made to the bones and the orientation of the limb was altered too. The humorous was lengthened and the shoulder joint was changed so that the humorous was swung out sideways, enabling it to move backwards and forwards in a horizontal plane as each stride took place, with a bending of the body in a sinuous motion reminiscent of the tetrapod's osteolepiform ancestry.

To cope with the weight of an elevated body, further anatomical changes had to occur. New bones developed in the hands and feet to become digits, which spread out to act as a weight-supporting surface. The limb girdles were modified also. In osteolepiform fish, the pectoral girdle was effectively part of the skull and the impact of the animal's weight from walking would be transmitted to the skull, so the pectoral girdle became separate in early tetrapods. Additionally, the pectoral girdle was modified, becoming more robust and firmly attached to the vertebral column in order to cope with the animal's weight.

Evolution also had to provide solutions to the issues of respiration and feeding. The complicated jaws, with various kinetic aspects, which fish evolved in the Devonian, were lost in the early amphibians. Their jaws were much simpler and lacked the mobility that fish had previously developed. Evolution appears to have started afresh; as the jaws adapted to feeding on small fish, but also terrestrial invertebrates hunted on land. However, amphibians retained the original teeth, which show labyrinthine internal patterns of infolding, similar to that of *Eusthenopteron*. This is largely because they continued to target the same prey species.

Respiration on land requires a supported vascular surface rather than gills. Modern lungfish have lungs, of course, and the same is assumed for the osteolepiformes. Consequently, the first amphibious tetrapods, presumably, also had lungs to extract oxygen from the air. At some point in evolution, a certain tissue must have become specialised as a gas exchange surface. Lungs contain many sacks, the alveoli, which increase the surface area for the diffusion of gases across the mucous-lined membrane to a vast network of capillaries, thus oxygenating the blood. The structure of the lungs of the early amphibians is likely to be much less intricate than those of the modern examples, though.

A familiar problem faced by life on land is the risk of desiccation when water is lost. Water can evaporate from the skin and from the lining of the mouth. As a consequence, amphibians are limited to fresh water environments, but the evolution of semipermeable skin coverings in some species reduced these restrictions.

Once amphibians became accustomed to life on land, their sensory systems had to improve in order to continue living a successful terrestrial lifestyle. As more tetrapods started to populate the land, competition between them would increase and the competitive environment was enhanced as the terrestrial arthropods responded to the new source of predation. The lateral line, while still being useful in water and retained by many aquatic amphibians, was obsolete on land. Eyesight would be important for identifying small prey on land and in shallow pools and, hence, a valuable asset. Hearing, however, was very limited in the air: the main bone associated with hearing in modern tetrapods, the stapes, is present in early amphibians, but is too large for the animals to be able to hear high frequency sound. Because sound travels better through water than air, the latter being the less dense medium, amphibians were only accustomed to auditory responses underwater, and would have to modify their physiology to improve this sensory system.

Amphibians evolved in response to this new environment and, as climate changed from the arid Devonian conditions to the carboniferous, tropical swampy conditions were created that suited this new class of animals and they were able to diversify greatly.

Carboniferous Life

The first tetrapods appeared in the late Devonian at a time when the climate was arid on land; only locations close to water sources, where plants could thrive, were conditions more amenable. It is here that the first tetrapods evolved from the osteolepiform fish. An example of one of the early tetrapods, generally considered to be among the oldest amphibians, is *Ichthyostega*. Almost complete remains of the 1 m long fossil of

Ichthyostega were found in the late Devonian of Greenland and provide an insight into the transition from lobe-finned fish to amphibians over the course of evolution.

Ichthyostega has an osteolepiform body outline, with deep vertebrae and a tail fin, but was obviously different from a fish because it had powerful limbs and had a shorter and broader skull, that is separate from the pectoral girdle. Additionally, the ribs are unusually large, having broad plate-like structures on the lower margins, which overlap to form an almost complete sidewall. This rib cage (or perhaps more accurately, rib box) may have provided some support for the internal organs, since *Ichthyostega* probably still had a weak osteolepiform-type backbone.

The appearance of tetrapods in the late Devonian was probably a result of unexploited resources found in the new terrestrial ecosystems and the intense competition in the aquatic environment, driving fish towards the land. However, lack of habitat in the arid Devonian climate probably restricted the evolution of amphibians, but that would change in the Carboniferous, when conditions changed in such a way that suited them: the climate became more moist and humid. A changing climate provided amphibians with the right conditions to undergo dramatic diversification; the Carboniferous, 360–286 Ma, was the time when the main phases of early amphibian evolution took place.

The changing conditions in the Carboniferous were the result of tectonic activity: the Earth's continents were coalescing into one land mass, which was situated largely along the equator. Within equatorial regions, tropical conditions prevail. Today these include the Amazon, Congo and Indonesia. During the Carboniferous, vast forests covered the ground; shrouded by mist produced by the transpiring vegetation. Damp forests evolved composed of gigantic trees and lush undergrowth of considerable proportions, including giant club mosses, 40 m tall lycopods, such as *Lepidodendron,* horsetails up to 15 m tall, for example *calimites*, which could live in 1 m depth of water. When the trees and undergrowth died, they would decompose in the anoxic conditions of the swampy ponds under the forest canopy to form peat, which eventually turned into coal through diagenesis. The decaying plant material on the forest floor provided a home for numerous ground-dwelling arthropods.

Arthropods, as well as amphibians, thrived in these swampy tropical conditions. The abundant flora expelled vast volumes of oxygen into the atmosphere, changing its entire composition. The oxygen-rich atmosphere suited the arthropods: it enabled them to assimilate more oxygen into their tissues and so were able to grow very large indeed. As well as attaining great size, during the Carboniferous, arthropods evolved the ability to fly.

The first animal pioneers on land were arthropods, such as millipedes, like *Pneumodesmus*, which had a body composed of many segments, with many feet. Some of these small arthropods found it to be an evolutionary advantage to reduce the number of segments and legs that make up their bodies because it enabled them to become more agile, hence more capable of escaping predatory amphibians and fellow arthropods. Eventually, the number of body segments was reduced to just three, which became a popular format and these were the first insects. They had a head, a thorax with six legs and an abdomen. Insects were agile and kinetic, and in the Carboniferous, some employed a new strategy to evade predation: the ability of flight.

As with all arthropods, insects moult their exoskeleton during ecdysis in order to grow. Unlike other arthropods, though, as they grow and develop, insects dramatically change their morphology and juveniles often lack wings that adult forms possess. Numerous insects live their juvenile lives in water and some will develop pre-wing structures, which they may use as paddles to propel themselves through the water. It is hypothesised that primitive aquatic insects that used similar paddle structures to move through water, evolved to become the first flying insects. The construct of wings from dead tissue, despite being fragile and unrepairable, was beneficial to the insects: exceedingly light-weight wings can vibrate rapidly to generate lift and provide insects with precise manoeuvrability. Insects are generally short-lived, so the risk of damage to the wings is not significant.

The attainment of large size became a trend for the arthropods in the Carboniferous, and this is true for some of the flying insects too. The largest flying insects known belong to the genus *Meganeura*. *Meganeura* were essentially giant dragonflies, with a wingspan of up to 75 cm. Details of the mouthparts reveal that *Meganeura* were predators. The feature that started off to avoid predation, then became a technique to hunt prey in the Carboniferous. *Meganeura* probably hunted ground-dwelling insects; catching them with its strong legs and devouring its meal while in flight. Due to its size, *Meganeura* may also have fed on small tetrapods.

Meganeura were by no means the largest arthropods to evolve in the Carboniferous; the small myriapods, that originally conquered land, evolved into a genera of enormous individuals, collectively known as *Arthropleura.* Their impressive size, some species growing to lengths of 2.5 m, is deduced from fossil evidence of tracks and sections of body armour. Unfortunately, the mouthparts are not preserved, so palaeontologists have to speculate what its lifestyle may have been: was it a docile plant-eater, like modern millipedes, or a voracious predator, as centipedes are today? However, evidence of plant spores found in parts of the fossil where the animals' guts are presumed to be, suggest that plants were at least part of the diet, but these may just be contaminants.

Arthropods could grow to such unprecedented sizes in the Carboniferous because the atmosphere was rich in oxygen due to the vast expanses of photosynthesising vegetation. Analysis of rock chemistry suggests that the atmosphere had an oxygen composition of 30–35%, compared to the 21% that it is today. With abundant oxygen in the air, a large concentration gradient is established along which the oxygen can passively diffuse into the respiring cells of the arthropods. Arthropods could grow so big because the plentiful oxygen entering through the spiracles could penetrate deeper into the tissue. The smaller differential between the concentration of oxygen in the atmosphere and the tissue of the insects means that shorter diffusion pathways are required; hence insects can't grow large today.

Besides the reduction in atmospheric concentration of oxygen, another crucial reason why arthropods have limitations regarding size is due to their chitinous exoskeleton. In addition to providing protection, arthropods rely on their exoskeleton for structural support: the external skeleton must cover the entire body and, as the animal grows larger, the exoskeleton must also grow larger and thicker, making it incredibly heavy. This great

weight also limits the size to which arthropods can gro
since the animal will be less nimble, it increases susce
predation, or the animal may even be unable to suppo
weight. Large arthropods were only a presence
Carboniferous because the vertebrates had not yet evol
larger animals that would readily predate the lur
behemoths, and atmospheric conditions ceased to be suit
enrich their tissues in oxygen as atmospheric levels
declined.

The evolution of flight in the Carboniferous meant
arthropods were the first animals to conquer the sea, land and
The giant arthropods that derived from the optimum conditi
in the Carboniferous forests are only a small chapter in the st
of arthropod evolution. Arthropods first evolved in the Cambri
540 million years ago, and during their history have endure
through every mass extinction that the planet has faced
Although global cooling and reduction in atmospheric levels of
oxygen meant that arthropods could no longer attain the
impressive sizes that they did in the Carboniferous, they
nevertheless continue to be perhaps the most spectacular group
of animals on Earth. The insects that evolved in the
Carboniferous became the most numerous and diverse animals
on the planet.

Vertebrates, having an internal skeleton and circulatory
system that can provide the tissues with the oxygen they need,
can grow large without physiological hindrances. The
Carboniferous was the time when terrestrial vertebrate evolution
really started to take-off: the amphibians began to diversify and
new evolutionary lineages arose. By this time, the amphibians
diverged into about twenty families, which are divided into three
prominent lineages. The labyrinthodontia have a large body size
and are characterised by the labyrinthodont tooth patterns, which
they share with osteolepiformes; the lepospondyli, which were
smaller; and the lissamphibia, comprising the modern groups,
such as frogs and salamanders.

To illustrate the diversity of the amphibians in the
Carboniferous, here are some examples. *Crassigyrinus* had a
deep skull and sharp teeth indicative of a carnivore; having
powerful jaws to seize struggling large fish. A slender body, with
reduced limbs, suggests that its hunting activities primarily took

...lace in water. *Proterogyrinus* was also a fish eater, but developed limbs indicate that it had terrestrial tendencies. However, a deep, flat-sided tail shows that it could swim competently. *Greererpeton* had an elongated skull, with eyes set forward and well-developed lateral line canals; it had short limbs and a long body; the neck and trunk being composed of 40 vertebrae, so presumably was aquatic. Members of the order Aistopoda were snake-like animals: they are presumed to have lost their limbs secondarily, rather than having evolved from limbless fish.

Towards the end of the Carboniferous and into the Permian, amphibians were predominantly terrestrial in habit, resembling lizards, with strong limbs and a robust skull designed to demolish the tough exoskeletons of terrestrial invertebrates, such as insects, spiders and millipedes. However, some did remain in the water, such as members of the order Necridea, for example *Diplocaulus*. *Diplocaulus* had a peculiar extended skull, shaped like an arrowhead. Biomechanical research suggests that this delta-wing shape provided lift to the animal as it strikes from a lurking position, obscured at the lakes bottom. *Diplocaulus* would have assaulted unwitting prey at the pond's edge; bursting out from the murky water to seize its unsuspecting prey.

The terrestrially-adapted amphibians that arose in the late Carboniferous and early Permian resemble reptiles, having strong limbs and more massive skeletons than their predecessors. An excellent example is *Eryops megacephalus*, which was a 2 m long carnivore and was among the largest vertebrate predators of the time, which fed on other tetrapods and occasionally fish, but its habits were predominantly terrestrial. *Eryops* belonged to the order Temnospondyli, which were common throughout the Carboniferous. Two other more advanced groups that appeared at this time, but were typically Permian, are the Seymouriamorpha and the Diadectomorpha; both of which are more analogous to reptiles than even *Eryops*; hence are generically termed reptilomorphs. *Semoria* was a 60 cm long land-living amphibian, found commonly in Midwest USA, and was different from the amphibians so far considered, in that its body was elevated higher off the ground. Its long limbs probably meant that it was capable of running quickly in pursuit of terrestrial prey. *Diadectes*, by contrast, is unique in that eight

peg-like teeth at the front of the mouth is evidence that it was among the first herbivorous terrestrial vertebrates. These teeth were used for nipping off portions of vegetation, and broad teeth further back acted as grinding platforms.

Derived from lobe-finned fish, early amphibians from the late Devonian continued to live predominantly an aquatic lifestyle. As evolution and adaptation persisted, amphibians were able to spend more time on land, where rich food sources, new habitat and low competition provided amphibians the opportunity to diversify throughout the Carboniferous. The amphibians came to occupy a variety of ecological niches presented by land in the swampy Carboniferous forests. Some continued to live in lacustrine open waters; some preferred to live closer to the shore; others moved from the lakes and occupied swampy ponds inland. Amphibians also found it advantageous to exhibit terrestrial habits: these groups eventually became the reptilomorphs. Amphibian evolution didn't end with the reptilomorphs, of course, as modern groups, the lissamphibia, became diverse; first appearing in the Mesozoic. They are represented by three distinct groups: the order Anura (frogs and toads), the order Urodela (newts and salamanders) and the order Gymnophiona (limbless Caecilians). Towards the end of the Carboniferous, though, certain clades of reptilomorph amphibians diverged to produce a new and discrete group of derived vertebrates: the appearance of reptiles.

The Appearance of Reptiles

The terrestrial landscape was dominated by temnospondyl and reptilomorph amphibians, particularly swampy forests, during the late Carboniferous. Among these organisms, the first true reptiles appeared; initially, small lizard-like creatures that fed on worms and insects; found in drier localities, where vegetation consisted predominantly of ferns. Such regions became increasingly common towards the climax of the Carboniferous: with more habitat, populations of small reptiles increased.

Perhaps the best example of early Carboniferous reptiles is the fossil remains of *Hylonomus lyelli*, which have been found preserved within the trunk of lycopod (club mosses) trees in several incidences. The unfortunate animals fell into the hollow of a rotten stump, inevitably died and became concealed following the subsequent infill of sediment. *Hylonomus* resembles modern insectivorous lizards and measures 20 cm in length. In contrast to many amphibians, the head is relatively small; approximately one-fifth the trunk length and is of light construction, which is indicative of the animal's food choice. Further anatomical aspects of the skull indicate that *Hylonomus* fed on invertebrates: small sharp teeth could readily pierce the tough exoskeletons and a new muscle group, the pterygoids, assist the adductors in pulling the jaws up, providing *Hylonomus* with greater crushing power than its amphibian counterparts.

The first reptiles were small because of thermoregulatory effects. Animals experience greater temperature extremes in air than water, which has a higher heat capacity, and so offers more continuity. In becoming more divorced from water, it was advantageous for the first reptiles to be small because they could take shelter more easily to avoid chilling or overheating; accommodating in among vegetation, cracks and crevices, or in hollowed tree trunks, whereas larger animals would find this difficult. Indeed, this is likely the reason why *Hylonomus* have been found preserved in fossilised tree trunks. Furthermore, smaller bodies are heated more quickly by basking, so the small, active lizard-like reptiles could readily overcome the morning lethargy and start hunting invertebrates on the forest floor. Hunting may also have occurred in the canopy of the forest, in an analogous way to salamanders today, given that their small body size would permit vertical climbing.

Progressive tectonic activity brought with it climate change as the continents drifted north of the equator. As the continents moved from the equator, they collided and merged together to form one giant landmass: the supercontinent Pangea. The arid sub-tropical conditions experienced were similar to the Devonian climate, except the expansive landmasses exacerbated the hot, dry conditions towards the centre of the continent, furthest from the moderating effects of the sea. This is evident from the extensive evaporate deposits from the Permian. The shifting climate resulted in floral changes: the swampy forests were lost, as the dominant lycopods and horsetails died out; replaced by hardy gymnosperms, such as conifers. With their habitat gone, amphibian communities regressed. The hostile Permian world provided reptiles with an opportunity: being better suited to the drier conditions, reptiles were able to replace amphibians as the dominant vertebrates on land.

Particular adaptations enabled the reptiles to surpass the amphibians in the Permian, at a time when climate change pressured life on land. A dry, scaly skin reduced water loss: a major problem in the desert environment. Amphibians, with their moist permeable skins, retreated to sparse lacustrine habitats, where they lived in high concentrations with intense competition. Reptiles, on the other hand, could occupy territory on land much more freely. Another advantageous characteristic was thoracic breathing, where muscles of the rib cage are used to aid the expansion of the lungs and facilitate breathing. A more mechanical approach was required to oxygenate the blood because atmospheric concentration of oxygen dwindled with the loss of the rainforest expanses and the prolonged volcanic activity, propagated as the continents collided, which released carbon dioxide and polluted the air. The most important evolutionary development, though, was the origination of the amniotic egg.

The amniotic egg enabled reptiles to become successful terrestrial organisms; as they became further divorced from water. Reptiles, unlike amphibians, don't have to lay their eggs in water and the aquatic larval stage, the tadpole, is omitted. For the first time in the history of life on Earth, vertebrates were able to breed on land. A shell protects the embryo and its food-source, the yolk, from the outside environment. By this method of reproduction, well-developed hatchlings are produced and the young have a greater chance of survival. However, the reptiles lay fewer eggs than either amphibians or fish, since amniotic eggs are far more expensive, in terms of energy resources, to produce.

The structure of the amniotic egg exemplifies the evolutionary adaptations required for reproduction to take place independently from water. There are two main features to the amniotic egg: a semipermeable shell and specialised extraembryonic membranes. The calcareous shell (although in snakes, some lizards and turtles this may be leathery) prevents the loss of fluids, therefore protects the embryo from exsiccation, but simultaneously permits the exchange of gases. There are several internal membranes: the chorion encloses the embryo and the yolk sack, while the amnion surrounds the embryo more closely. Both offer additional protection, compartmentalise the egg and permit gas transfer: oxygen needed for respiration can enter, while waste carbon dioxide can be expelled. The allantois also functions in respiration and forms a capsule for the storage of waste material. As the embryo develops, the proteinaceous food source, the yolk, is depleted and the allantois fills up.

As the reptiles radiated in the Permian, different groups became established, which can be distinguished through observation of the temporal fenestrae. The temporal fenestrae are openings behind the orbit (eye socket); the purpose of which is to reduce the weight of the skull and conserve calcium. Bone is heavy and expensive to maintain, so it can be advantageous to dispense with it where possible. Most areas of the skull are under stress from the jaws and neck muscles, but some regions are not subjected to stress; here cavities may develop without detriment to the effectiveness of the skull. Additionally, the temporal fenestrae provide edges for the attachment of specific muscles.

On the basis of temporal fenestrae, four groups of reptiles (and indeed other amniotes) can be acknowledged. A) Anapsid: reptiles with no temporal fenestrae, including early forms, such as *Hylonomus*, and the turtles. B) Synapsid: mammal-like reptiles and mammals, with one lower temporal fenestra. C) Diapsid: includes most reptiles and birds, with two temporal fenestrae. D) Euryapsid: marine reptiles of the Mesozoic, with one upper temporal fenestra, thought to be derived from the loss of the lower temporal fenestra in diapsid ancestors.

The most prominent group of reptiles from the early Permian, 286 Ma, were the pelycosaurs, representing 70% of all genera. These synapsid reptiles include a conspicuous clade of "sail backs". One example is *Dimetrodon*; a 3 m long carnivore. *Dimetrodon* had a thick skull, providing a powerful bite, and sharp teeth to pierce flesh. This robust meat-eater would have preyed upon other pelycosaurs that were herbivorous.

The function of the neutral spines, which radiate from the vertebrae to form the sail, may relate to a basic biological principal: the surface area-to-volume ratio of an animal in relation to heat exchange. The growth of the spines seems to be in relation to the weight of the animal, rather than length. As the animal matures and attains greater bulk, its body volume increases. In response, the neutral spines elongate and the surface area of the animal increases. As any animal grows, its volume increases roughly in proportion to the cube of its length, while its surface area only increases in proportion to the square. Adaptations to increase the surface area of an animal's skin relative to its volume are often derived from the physiological implications of homeostasis; specifically heat balance.

The neutral spines contain grooves that were probably occupied by blood vessels: the heavily vascularised skin of the sail likely had thermoregulatory purposes. As with reptiles today, the Permian synapsids, such as *Dimetrodon*, were most probably ectotherms; absorbing heat from the external environment, rather than generating body heat through metabolism of food. In the early morning, *Dimetrodon* would still have a low body temperature from the previous cold night, and consequently be sluggishly inactive. Having a broad sail though, *Dimetrodon* would readily warm its body by basking in the sun. It has been calculated that a large 250 kg *Dimetrodon* would require 12 hours of basking to raise its body temperature from 25 °C to 35 °C if it lacked the sail. With the sail, this would only take three hours, due to the additional surface area. As a predator, this gave *Dimetrodon* a selective advantage: it could attack its still torpid prey while they were basking. However, other pelycosaurs and their contemporaries, which lacked sails, were still successful in the early Permian.

Another synapsid clade, the therapsids, overtook the pelycosaurs by the mid-Permian and became very diverse. Therapsids encompassed both carnivorous and herbivorous lineages. The most dominant carnivores, by the latter end of the period, were the gorgonopsians, consisting of 35 genera; all of which are anatomically similar. One example is *Arctognathus*; a 1 m long predator being of typical size among gorgonopsians. It was capable of opening its jaws to a 90° gape, thus exposing its elongate canines, enabling *Arctognathus* to pierce the thick skins of the large herbivores that also evolved at this time, such as the dicynodonts. The dicynodonts were the dominant herbivores during the late Permian, comprising 70 genera. Some genera attained large sizes of up to 3 m or so in length. Presumably, these animals had the ecological role analogous to the grazing animals of today. One example is *Kannemeyeria*: a stocky animal with a robust skeletal structure. The success of the dicynodonts, such as *Kannemeyeria*, in the late Permian probably relates to their specialised jaw apparatus. *Kannemeyeria* had a horny beak, which it used like pinchers to strip off vegetation, which would be passed to the cheek region to be ground up by the teeth before being swallowed. Another aspect of the feeding apparatus is the tusks, used to scrape up roots, horsetail stems and ferns. Another therapsid group from the latter end of the Permian are the cynodonts, which include the ancestors of the mammals. These reptiles have many mammalian characteristics: modified jaw anatomy enables the attachment of greater muscle mass and elongated nasal bones fashion a skull morphology that resembles those of mammals. Cynodonts were small animals that generally fed on insects.

A diversity of non-synapsid reptiles also radiated in the Permian. A dominant group that appeared in the late Permian were the diapsids. These animals originated as small insectivorous lizard-like reptiles. Most were conventionally lizard-like tetrapods, such as *Youngina*, but more exotic forms existed also. For example *Coelurosauravus* had an extended rib cage, which, covered with a skin membrane, enabled it to glide from tree-to-tree as the living lizard Draco does. The ribs were retractile, so could be folded back when the animal was running. The descendants of these initial diapsid reptiles became divided into two discrete groups after the Permian: one included the dinosaurs, crocodiles and birds, the other led to the lizards and snakes. The two groups are called the archosauromorpha and the lepidosuromorpha, respectively.

By the end of the Phanerozoic, the Earth was set for the proliferation of reptilian groups which would dominate life on land throughout the Mesozoic. They rallied through catastrophic climatic events that marked the interface between these two eras and became highly successful and diverse.

Mass Extinction

247 million years ago, the most severe extinction in the history of life on Earth took place. This catastrophic event marks

the boundary between the Permian and Triassic periods, occurring over a time range of about ten million years; ending abruptly at the climax of the Palaeozoic. Perplexingly, this event is unassertively referred to as the Permo-Triassic extinction; a belittling title for the extreme annihilation that life was subjected to at the end of the Palaeozoic. Conditions on Earth deteriorated due to several factors, which together made the planet almost inhospitable. Some of these factors are apparently interlinked, while others are separate incidences. Initially, the jeopardy faced by life took a gradual approach, through plate tectonics and marine regressions. As conditions worsened in the last 150 thousand years of the Permian, volcanism occurred, which has been attributed to the abrupt demise of much of the life on Earth.

The extent of the extinction was colossal and its causation coincides with the loss of two major structural ecosystems: coral reefs and forests. The marine environment suffered the most, with 50% of marine invertebrates dying out, including hitherto successful groups, such as the trilobites and eurypterids. On land, of the 37 tetrapod families in existence during the last five million years of the Permian, 27 died out (a loss of 73%). Most tree species became extinct, along with a lot of other flora. In total, it is estimated that an incredible 90% of all species became extinct at the end of the Permian.

Paradoxically, the formation of the supercontinent Pangea, which provided reptiles with the conditions necessary for them to flourish, also contributed towards many species, among which reptiles were included, dying out. With all the land concentrated in one continent, the relative amount of coastal lowland was reduced; therefore, so was the variety of habitats. Life on land had a limited number of niches available, so the potential for diversification during the onset of climatic upheaval was restricted: life on land was unable to adapt to the changing conditions. Ironically, the arid conditions that enabled the reptiles to triumph over amphibians undermined their capacity for survival as conditions became drier. Terrestrial life was orientated predominantly towards the forest ecosystem, which, in the adverse conditions, became undermined.

The formation of Pangea also meant that there was a loss of continental shelf environment. Coral reefs are very delicate, requiring particular conditions, namely warm shallow waters in which light can penetrate. Coral reefs are locations of high biodiversity, so the disruption caused by changing environmental conditions would have a very negative impact on Permian marine faunas. Pangea restricted the available continental shelf environment in which reefs may thrive; this predicament was extenuated when an immense marine regression occurred towards the end of the Permian. Global cooling caused sea water to be stored in glaciers located in the southern hemisphere, evidence for this comes from the evaporate deposits behind the fossilised coral reefs in the ancient Zechstein Sea of NW Europe. As the shoreline regressed, episodic flooding, followed by drying out, created hypersaline waters, which left a salt deposit. Consequential to this devastation, the dominant reef builders, rugose and tabulate corals, disappeared entirely. It would be 15 million years before coral reefs were built once more.

Pangea and marine regression presented a gradual approach to the deterioration of conditions for life in the Permian, but a more abrupt catastrophism provided a "sting in the tail" for the end Permian extinction. Extensive volcanism occurred in Siberia 252 Ma from enormous rifts in the Earth's crust; extruding basalt lava, which produced a vast formation known as the Siberian Traps. It is thought that the volcanic activity was caused by a very large mantle plume; a hot region of magma under the crust. Eruptions led to flood basalt over a region larger than Europe. The extensive lava formations would have had a drastic localised effect on life, but more cataclysmic were the emissions of large volumes of gas. The large clouds of sulphate released would have produced a haze that would reflect the sunlight, causing an episodic cooling of the planet for a short while. This effect was soon relinquished because the haze dissipated as droplets of precipitation fell from the sky as sulphuric acid rain, poisoning the ground. More long-lasting was the global warming that followed from the greenhouse effect, resulting from the copious amount of carbon dioxide escaping to the atmosphere, which was also released from the volcanoes in abundance.

Pulses of flash warming following the cold snap would generate a fluctuating climate that life on land would struggle to cope with. The lack of sunlight and unrelenting acid rain would have devastated the forest ecosystems: without food, animals perished. The high concentrations of carbon dioxide in the atmosphere resulted in CO_2 being dissolved in the ocean. High global temperatures and all the land accumulated in one continent, which would reduce the turbidity of ocean currents, caused the ocean to stagnate and CO_2 accumulated in high concentrations in the water; it acidified the blood of marine organisms, poisoning them, and eventually led to widespread extinction.

Fossiliferous limestone deposited at the end of the Permian show a radical change in their chemical composition around the time of the extinction. Limestone is made from microscopic creatures (coccoliths) which build their skeletons by combining calcium and carbon dioxide to form calcium carbonate. The composition of the limestone indicates a global disaster: it is composed of proportionally much more of the ^{13}C isotope than usual. Plants, preferentially, take up ^{13}C from the atmosphere for photosynthesis, thereby reducing its abundance in the environment, so the coccoliths build their skeletons out of the more common ^{12}C isotope, therefore limestone is composed mostly of ^{12}C. However, the limestone at the P-T boundary contains a much higher percentage of ^{13}C. This is because the destruction of plant life meant that the isotope with organic affinities, ^{13}C, was less frequently assimilated by plants and locked up, and decaying organic material produced CO_2 rich in ^{13}C, hence coccoliths incorporated more of this into their skeletons and this chemical property of the limestone is indicative of the catastrophic events endured by life at this time.

As concentrations of carbon dioxide gas increased, conditions in the stagnant oceans became anoxic: this led to the formation of black shale, rich in pyrites, with organic derivations. After the abundant fossils in rocks from during the extinction, with life on Earth depleted, very few fossils appear in records just after the Permian and, with loss of forests, a coal gap arises in the Triassic. This concludes the main points of evidence for the P-T extinction.

Mass extinction may be a melancholy portrayal of the events experienced by life under unforgiving conditions, but paradoxically the eradication of some species is an essential feature for evolutionary progression and deriving new species. The history of life on Earth has been punctuated by several mass extinctions, where 40–50% of species die out in a relatively short time, but none compare to the magnitude of the P-T event. Extinction is a normal part of evolution and does not occur exclusively in large-scale catastrophism. In fact, most species become extinct due to localised effects, such as the drying out of a river, flooding of an island, volcanic eruption or simply out-competed by superior species; events which occur much more frequently than a mass extinction. This is known as the background rate of extinction. Remarkably, it is estimated that 99.9% of species that have ever existed are now extinct.

Background extinction and mass extinction contrast as two fundamental aspects of evolution: gradualism versus catastrophism, respectively. A community of any scale remains in equilibrium, provided conditions are not changed substantially. For the majority of the time, an ecosystem will remain static in this way: a few new species will evolve through the slow process of natural selection, as contemplated by Darwin, and will out-compete indigenous populations. The continual loss of some species and replacement by new species persists in this way, leading to a slow rate of evolution. Life could not have achieved its advanced status today if it wasn't for catastrophes. The diversity and complexity of life observed at present would not have occurred through gradual evolution. A catastrophic event that causes mass extinction by eradicating many species in a short period of time dramatically disturbs equilibrium. Suddenly, a variety of ecological niches becomes available after the extinction event. Natural selection is accelerated as species adapt to a particular niche, free from the inhibition that arises through competition. Soon, an entirely new ecosystem is derived, filled with new species, and once again life attains equilibrium.

Through the course of the history of life on Earth, a punctuated equilibrium has arisen from the around half dozen mass extinction events that have occurred. These have been frequent enough to interrupt the Darwinian gradualism and accelerate evolution to the extent that the remarkable lifeforms observed today evolved. The P-T extinction though, dwarfs the others, caused as conditions for life on Earth became very poor indeed. With the formation of Pangea, marine regression and volcanism occurring simultaneously, life barely survived, but despite the devastation, mass extinction is a vital part of evolution.

The Dinosaur Age
Origin of the Dinosaurs

The Triassic period, 245 million years ago, was much the same as the Permian: the continents remained united in Pangea; climatic conditions continued to be homogenously hot and dry, and reptiles were the dominant land vertebrates; albeit depleted in numbers and diversity following the great mass extinction. The mammal-like synapsids survived the catastrophic event, but lost many adaptive zones to a new group of diapsid reptiles: the archosauromorpha. Vacant ecological niches arose due to the loss of many species during the great mass extinction and this provided members of this clade with the opportunity to radiate into several new groups. Among these were the dinosaurs.

The most important group of the archosaur branch is the archosauria. Most of the Triassic archosaurs were thecodontians, which are the ancestors of crocodiles, pterosaurs and dinosaurs. Early Triassic thecodontians capitalised on the carnivorous niches bequeathed to them by the gorgonopsians, which died out at the P-T boundary. *Proterosuchus*, from South Africa, possesses three features that are typical of an archosaur: a gap in the side of the skull, between the eye socket and the nostril, called the antorbital fenestra; flattened (rather than rounded) teeth; and finally, a knob-like structure on the femur for muscle attachment, called the fourth trochanter. *Proterosuchus* was a slender carnivore, measuring 1.5 m in length. It would have preyed upon small to medium-sized reptiles, such as the dicynodonts and cynodonts. Although *Proterosuchus* was an advanced reptile in many respects, it still retained the primitive format of four quadrupedal limbs positioned in a sprawling posture, as preserved in modern salamanders and lizards.

Another South African thecodontian from the early Triassic, called *Euparkeria*, had much more advanced characteristics. Despite being small, only 0.5 m in length, *Euparkeria* would have been an effective predator. It had long limbs, with a reduced and flexible ankle arranged in an erect posture, so it may have been capable of walking on all fours and also bipedal. Running on two long hind legs would have enabled *Euparkeria* to swiftly capture prey, thereby omitting the necessity of expending excessive energy during long-distance pursuits and also increased the frequency of successful hunts.

By the middle of the Triassic, evolutionary diversification had divided the archosaurs into two major groups: the crocodylotarsi and the ornithosuchia. The former includes ancestors of the crocodilians; the latter includes dinosaurs, pterosaurs and, ultimately, the birds. The crocodylotarsi had advanced features in common with *Euparkeria*, which pterosaurs lacked. The limbs were in a semi-erect gait and intricate ankle bones enabled this group to walk and run more efficiently because there was more fluidity in the movement.

The primitive crocodylotarsans existed in great diversity in the mid-Triassic. *Parasuchus* resembled modern crocodilians; being adapted for fish eating. Its nostrils were positioned on a ridge in front of the eyes (rather than at the end of the skull, as with crocodilians today), which enabled it to submerge itself in the water, remaining hidden before lurching forward and grabbing unassuming tetrapods from the water's edge. The struggling prey were held fast in the long rows of interlocking teeth. Other crocodylotarsans evolved to become the major predators at the time. *Saurosuchus* grew to 7 m in length and its limbs were positioned under the body like pillars to enable a smoother running motion, and could rapidly dispatch a dicynodont. Some crocodylotarsans even evolved to become the first herbivorous archosaurs. *Stagonolepis* had a thick-set 2.5 m long body and a shovel-like snout, presumably for digging up roots.

Alongside the archosaurs, other groups of archosauromorph diapsid also existed. For example, the herbivorous rhynchosaurs, such as *Hyperodapedon*, which fed on tough plant matter, including ferns. They would have grazed these plants that grew around water sources, such as lakes, rivers and episodic locations like player lakes and wadi channels. *Hyperodapedon* was adapted with the necessary apparatus for dealing with this tough vegetation. A very broad skull, which was in fact greater in width than in length, provided room for strong muscles that closed the jaw via a simple pivot mechanism. This jaw action lacked any forwards and backwards or side-to-side movement and is termed the precision shear system; cutting through tough plant material like a pair of scissors. *Hyperodapedon* also had massive claws on the back feet, used for digging up roots and tubers.

Adaptation of reptiles to adopt an herbivorous lifestyle produced two solutions regarding physiological prerequisites necessary to live on vegetation. Because tough plant material is difficult to digest, vertebrates must enlist the help of fermenting bacteria to break down the cellulose, and, as a result, need large guts to serve as vats in order to process large amounts of the protein-poor foodstuff. The first solution, therefore, is large size. Alternatively, vertebrates can selectively feed on the nourishing parts of plants; reproductive parts tend to be more proteinaceous and sugary. This second solution requires vertebrates to be small, with dextrous mouths. Adaptation towards herbivory in reptiles may have started in the Permian-Triassic, but these adaptations were most spectacularly illustrated by the dinosaur faunas.

An excellent case study that documents the transition from a landscape dominated by early reptiles, namely thecodontians and gorgonopsians, to the new dinosaur faunas is the Lossiemouth Sandstone Formation from Elgin in North-East Scotland. The fossils date from the late Triassic, 227 Ma, and are preserved in dune deposits; evident from the characteristic cross-bedding and well-rounded grains. Sandstone is very hard due to the quartz composition, so the bones are difficult to extract. Palaeontologists employed a technique where acid is used to dissolve the bone and a flexible rubber/plastic cast is made from the resulting mould, providing exquisite detail of the bone. The Elgin reptiles include a community of herbivorous genera, among which *Hyperodapedon* is present. These animals would have frequented water sources amid the arid desert environment, feeding on tough plants, which they dug up with their snouts and claws. They would have been preyed upon by an advanced thecodontian predator of the time, *Ornithosuchus*, which was a precursor to the new dinosaur genera.

The thecodontians underwent remarkable diversification during the Triassic. The crocodylotarsans clade produced a number of 'crocodilomorphs', which, although are conspicuously different to the true crocodilians that evolved in the Jurassic; being fully terrestrial, insectivorous and bipedal, shared a number of anatomical similarities. The ornithosuchan clade radiated into advanced forms, such as *Ornithosuchus*, from which the dinosaurs are derived. *Ornithosuchus*, having a slender body, powerful skull and being intermittently bipedal, was very dinosaur-like and filled the ecological niche that dinosaurs would later takeover.

The Triassic takeover, where archosaurs replaced the therapsids could have been due to a radiation after the therapsids had largely gone, due to climatic reasons causing extinction. However, there is strong evidence for direct competition, where the carnivorous archosaurs out-competed the carnivorous therapsid, such as the gorgonopsians, and hunted out the remaining herbivorous therapsids, such as the dicynodonts. Herbivorous archosaurs then later evolved to take advantage of the abundant plant life in the now vacant niches. This model implies that the archosaurs success was due to competitive superiority. Evidence can be assembled that shows how diapsid reptiles, particularly the archosaurs, were superior to their synapsid contemporaries in two key areas: respiration and locomotion. From this, palaeontologists can infer how their metabolic and thermoregulatory traits were advanced over the primitive therapsids. This evidence offers profound reasoning to explain the triumph of archosaurs and the decline of synapsids, except for the resilient, discerning ancestors of the mammals, during the Triassic.

The relationship between respiration, locomotion and physiology has a fundamental floor in the primitive reptiles. Since the evolution of lungs in osteolepiform fish, respiration and locomotion have had a superficial incompatibility. The sprawling gait of amphibians, and later the synapsids as well as modern reptiles, which is a vestige of sinuous swimming motion, causes the trunk to twist from one side to the other when walking. This results in one lung being compressed in the thorax, while the other expands, alternately, with each step. This cycle obstructs and essentially prevents normal breathing, where the chest cavity and both lungs expand and contract uniformly. Basically, these animals are not physiologically adapted to breathe and locomote simultaneously. They rely on anaerobic glycolysis, which causes an oxygen debt, so high performance cannot be sustained. The therapsids could run for short periods, then would have to stop and stand with their feet placed symmetrically, in order to catch their breath. Hence, ambush tactics were the predominant hunting strategy for these animals. The synapsid solution, observed in pelycosaurs, was to stiffen the backbone to restrict sideways movement, as well as to have shorter limbs and a longer body, so each step was smaller. However, this only partly reduced the constraint and impeded on locomotion.

The diapsids were superiorly adapted. The archosaurs had adapted biomechanics of locomotion by evolving an erect gait. The trunk was supported by limbs underneath the body, so that the body didn't twist as the animal ran. Moreover, this physiology encouraged breathing on the run because the backbone flexes and straightens in the vertical plane, synchronised with the contractions of the diaphragm to alternately expand and compress the chest cavity evenly, facilitating breathing. The consequences of this are that sustained running can be achieved because the muscles are constantly supplied with oxygen to mitigate fatigue, so thecodontians were able to run down and destroy their synapsid contemporaries.

By the late Triassic, 225 million years ago, dinosaurs started to appear. Initially, they were only rare, but radiated towards the end of the Triassic. The oldest dinosaurs were generally bipedal carnivores, similar to their ornithosuchan ancestors. *Herrerasaurus*, from Argentina, was 3 m long, heavily built and an active hunter; primarily targeting small tetrapods. The hip and leg bones of *Herrerasaurus* were specifically adapted for bipedalism, and thus surpassed the laboured bipedal locomotion of the less advanced thecodontians. Unlike some thecodontians, where the limbs were fitted under the pelvis like pillars, bipedal dinosaurs, such as *Herrerasaurus*, had an inwards bend of the head of the femur, so that it fitted sideways into the acetabulum opening in the hip.

Another early dinosaur is *Coelophysis* from North America. It too, was a fully-bipedal carnivore, similar in length to

Herrerasaurus, but was of very lightweight construction, probably weighing around 25 kg. With the modified hip arrangement, *Coelophysis* could run very fast, with great agility, enabling it to capture swiftly moving lizards.

Towards the end of the Triassic, dinosaurs had diversified greatly and had become the most abundant vertebrates on land. At this stage, herbivorous varieties had evolved; the best known of which is *Plateosaurus*. *Plateosaurus* is known from over 50 locations in Germany and Switzerland. Like its thecodontian ancestors, *Plateosaurus* could exhibit either a bipedal or quadrupedal posture, but retained the habitual body proportions of a fully-bipedal dinosaur. The ratio of forelimb length-to-hind limb length is equal to one half, that is, the hind limbs are twice as long as the forelimbs, characteristic of bipedal tendencies. *Plateosaurus* had a long neck and a long tail, which provided balance, and may have adopted a quadrupedal posture when grazing ground level vegetation. Permanent quadrupedal posture was not a necessity for supporting body weight, as is the case in later long-necked dinosaurs, so the option of bipedal locomotion meant that it could speedily flee from predators. It had serrated teeth with which it cropped the plant material before swallowing it whole. This brought the necessity for *Plateosaurus* to swallow pebbles to serve as gastroliths for grinding the plant material into a digestible state within the gastrointestinal tract.

By the beginning of the Jurassic, dinosaurs had become the most abundant terrestrial vertebrates. There are currently two theories to explain the radiation of the dinosaurs in the late Triassic. It is generally considered among palaeontologists that dinosaurs either radiated opportunistically after the mass extinction, in which other reptile groups died out, or they competed with the synapsids, the rhynchosaurs and the thecodontians over a longer time range and, eventually, prevailed.

Competition within and between species results in the slow process of natural selection and eventually the genesis of new groups of organisms. The origin of the dinosaurs may be traced back to the mid-Triassic, where they may be considered as derivatives of the archosaurs. Reptiles with superior mechanisms of locomotion had the selective advantage and prevailed; causing the extinction of their 'less-well-adapted' contemporaries. In evolutionary terms, the dinosaurs are regarded as a great success: can this be due only to the fundamental and axiomatic law of natural selection? The alternative theory describes the dinosaurs as 'making the most' of an extinction event that removed the synapsids, thecodontians and rhynchosaurs and provided the dinosaurs with the chance to occupy the vacant ecological niches. Conveniently, the dramatic radiation of the dinosaurs at the end of the Triassic coincides with another small extinction event. The bipedal, agile carnivorous dinosaurs, such as *Herrerasaurus* and *Coelophysis*, would have been better adapted for a scavenging lifestyle than the cumbersome quadrupedal thecodontians.

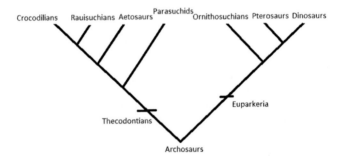

Figure 5: Cladogram showing the divergence of the archosaurs clade to illustrate the relationship with other reptiles, including the origin of the dinosaurs.

The Dinosaurs

Dinosaurs reached their acme during the Mesozoic; they diversified and increased in numbers at the beginning of the Jurassic and continued to be the most abundant group of land vertebrates through the Cretaceous. They dominated the land for 160 million years and evolved into numerous forms; adapting to fit all the ecological niches presented to them by the Mesozoic world. Bipedal carnivores persisted long after the Triassic; being the apex predators for the entirety of the dinosaur age. Moving on from the first examples, such as *Herrerasaurus*, they became very successful and diverse. Meanwhile, herbivores evolved into long-necked advanced forms similar to *Plateosaurus*, armoured quadrupeds, and agile bipedal forms also.

The dinosaurs are divided into two principal groups: the saurischians (lizard hip) and the ornithischians (bird hip). As the names suggest, these two groups differ by the morphology of their pelvis. The more primitive structure of the saurischians has the pubis pointing forward and the ischium back, as in the basal archosaurs of the Triassic. The ornithischians, by contrast, have the pubis pointing backwards parallel to the ischium and have developed a prepubotic process in front.

The saurischian dinosaurs include the early forms, such as *Herrerasaurus, Plateosaurus* etc., as well as their descendants. The bipedal carnivores belong to the sub-order of theropods, while quadrupedal plant-eating 'long-necks' make up the sub-order sauropodomorpha. The theropods include the basal group, called ceratosaurs, which reached body lengths of 5–7 m, and the more advanced giant meat-eaters, the carnosaurs. The sauropodomorpha clade is comprised of the late Triassic/ early Jurassic prosauropods and the larger, more advanced, sauropods, which are direct descendants. The ornithischian dinosaurs were all herbivores and were comprised of two main groups; namely cerapoda and the thyreophora.

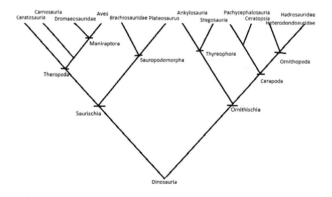

Figure 6: Cladogram showing the relationships between the dinosaurs.

By the middle of the Juassic, the supercontinent pangea began to break up, forming an opening filled by the Tethys Ocean. This created a greater area of coastal lowlands and the high sea level of the time flooded the lowest area, forming a lacustrine environment. Further inland, subtropical conditons prevailed, but with more moisture than in the Triassic. The dominant floras, including ferns and conifers of subtropical varieties.

In the early-to-middle Jurassic, carnivorous dinosaurs continued to be the typical ceratosaur shape: lightly-built, agile predators, with long limbs and tails. One of the earliest ceratosurs was *Dilophosaurus*. The skull of *Dilophosaurus*, to which it owes the name two-ridged lizard, is curious due to the two bony crests on the roof, which presumably were covered with skin and maybe had a signalling function. The skull was not particularly robust; being of light construction and had a weak joint between the maxilla and premaxilla, so the 7 m long *Dilophosaurus* probably preyed on smaller ornithopods in the scrubland. Later certosaurs, such as *Ceratosaurus*, had stronger skulls and heavier teeth adapted to tackle large prosauropods, such as *Thecodontosaurus*.

The ornithischian dinosaurs appeared around the same time as the saurischians. *Pisanosaurus* from the late Triassic of Argentina may have been the first ornithischian, but is only known from incomplete specimens. The cheek teeth have a triangular crown and are set back from the edge of the jaw, leaving a shelf, which indicates that *Pisanosaurus* had cheek pouches, where a layer of skin covers the face in order to retain food while other foodstuff is being processed. This is typical of an ornithischian. More complete skeletons of *Lesothosaurus* (also called *Fabrosaurus*) have been found in southern Africa from the early Jurassic and is more typical of an early ornithischian dinosaur. The skull contains two new bones: the predentory at the tip of the lower jaw and the palpebral contained within the orbit. The "bird hip" showing the ischia and pubis pointing backwards, and are fused at their tips. *Lesothosaurus* was a small, 0.9 m long herbivore; a likely source of prey for *Dilophosaurus.*

Living at the same time as *Lesothosaurus*, 208 million years ago, *Heterodontosaurus* was among the first ornithopods. The ornithopods were the largest and most successful group of ornithischian dinosaurs, which became particularly significant during the Cretaceous period, when they formed herds of large herbivores. *Heterodontosaurus* though, was a small dinosaur, similar in appearance to *Lesothosaurus*. However, a conspicuous difference between the two are the teeth. *Heterodontosaurus* (literally meaning "different-tooth reptile") had differentiated teeth, while *Lesothosaurus* had uniform teeth, with symmetrical-wear facets either side of the pointed crown; indicative of the mechanical up-and-down motion of the jaw, with no possibility for sideways movement.

The mouth of *Heterodontosaurus* contains two incisors, a canine and twelve cheek teeth. The bottom canine fits into a notch in the upper jaw. Since *Heterodontosaurus* was a herbivore, it is believed that the canines, which are omitted in some specimens, were a secondary sexual characteristic of males, used for defence and social display. The jaw hinge permitted movement of the lower jaw by rotation about its long axis. This enabled the cheek teeth to have a shearing effect. In later ornithopods, movement in the maxilla, rather than the lower jaw, had this shearing effect and is believed to be one reason why they became so successful.

Small bipedal herbivores, such as *Lesothosaurus* and *Heterodontosaurus*, may have sought protection when feeding in open scrubland by associating with the larger prosauropods, which were too massive for *Dilophosaurus* to attack. Using large body size as a defence against predation caused the direct descendants of the prosauropods, the sauropods, to become the biggest terrestrial animal ever to exist on Earth.

Characterised by their long necks, the sauropods from the mid-Jurassic were notorious giants. The long neck, which was balanced with a long tail, enabled the sauropods to reach into the canopy and graze the leaves of conifers and cycad trees; another advantage of being large. The legs were adapted to weight-bearing: they were relatively short and massive, with wide feet placed under the body in a pillar-like arrangement, as in elephants. Sauropods would have to spend most of their time feeding in order to support such a huge body size on low quality food.

The camarasaurids and the brachiosaurids share similar characteristics of the skull, where a curved internarial bar separates the nostrils on the high-domed skull. This enabled the suropods to breathe freely when grazing; as nostrils at the end of the snout would be conjested in the leaves. The diplodocids and titanosaurids had longer snouts, with a small number of peg-like teeth for cropping vegetation. Camarasaurids would have grazed the tough leaves of the shrubby trees that grew on flood plains. A huge, 18 m long, sturdy body would have contained large amounts of intestines in order to digest the tough leaves. Swallowing gastroliths would have helped grind the matter in the gizzard. By eating the more abrasive leaves, *Camarasurus* could live alongside other sauropods, such as *Diplodocus*, which fed on softer vegetation, without being in direct competition for food. However, young *Camarasurus* appeared to have eaten softer vegetation, from the signs of wear on the teeth. Perhaps their guts lacked the propensity to digest the tougher material when the animals were not fully grown.

To mitigate their colossal weight, the suaropods had hollow vertebrae, containing a complex of cavities that were supported by the lattice work of fine bone struts. The femurs of Brachiosaurus were massively thick to support the great weight. Unusually for sauropods, the upper arm bones of *Brachiosaurus* were longer than the femur (hence the name "arm lizard") and a long neck consisting of 13 vertebrae, meant that *Brachiosaurus* could reach leaves that no other grazing animal in the open woodland could access; up to heights of 16 m. Due to its huge frame, *Brachiosaurus* is estimated to have weighed in excess of 55 tonnes.

The skull of *Brachiosaurus* had a very small braincase. A tiny brain would only have been able to coordinate locomotion and anatomical functions, having very little individual thought capacity. *Brachiosaurus* would have operated in small herds focused on feeding. Their great size meant that consideration of predaors wasn't necessary. Supplying the brain with blood would have been difficult; a strong heart muscle would have forced blood at high pressure up the arteries in the neck, these arteries contained valves to prevent the blood from falling back down and were thick-walled and elastic to withstand the high pressure.

In the mid-Jurassic, 150 million year ago, dinosaurs grew big to combat predation. In addition to size, some ornithischians, such as the thyreophorans, developed armour defences. *Stegosaurus*, from North America, had two pairs of 1 m long spikes at the tip of its tail. By lashing the tail, a 9 m long *Stegosaurus* could inflict a fatal stabbing blow to a predator. As well as spikes, *Stegosaurus* had an array of plates along the ridge of its back. However, palaeontologists no longer consider these plates to be a defensive armour. The body plates developed independently from the skeleton within the skin and didn't meet the bones of the spinal column at all. They must have simply been rooted in the muscles of the back and held fast by massive ligaments; not robust enough to repel an attack from a carnosaur. Modern theories postulate that the plates may have been display structures or thermoregulatory devices. Grooves in the plates probably contained branching networks of capillaries during its life. By pumping bloood into the plates, the skin colour would have changed and may have acted as a courtship display, or perhaps a deterant to predators. It has been noted that the arrangement of the 80 cm long plates mirror the engineering design model for heat dissipation devices. Similar to the sails of the pelycosaurs, *Stegosaurus* may have altered its orientation to the wind; an overheated *Stegosaurus* could readily cool down by directing blood to the plates and standing broadside to the prevailing wind.

As many herbivorous dinosaurs increased in size during the latter stage of the Jurassic, so did the theropod predators that hunted them: thus evolved the large meat-eating carnosaurs. At 12 m in length and weighing around 3 tonnes, *Allosaurus* was one of the main predators of the giant herbivores, hunting young Brachiosaurs and Diplodocids, as well as thyreophorans like *Stegosaurus,* in the open woodland. At Cleveland-Lloyd Dinosaur Quarry in Utah, USA, the remains of 60 Allosaurs were found. The dinosaurs perished after becoming trapped in the wet mud surrounding a watering hole, after being lured by already trapped herbivores.

The large sauropods of the Jurassic exemplify one conspicuous trait of the dinosaurs: evolution of large size. The dinosaurs were a hugely diverse group of animals, varying in size from the chicken-sized *Compsognathus*, to the gigantic sauropods, such as the 80 tonnes *Brachiosaurus*. Large size is a conspicuous charcteristic for which dinosaurs have become renowned; being the largest land animals ever to have existed. Understanding how these giant animals existed is a tantalising aspect of palaeontology becaue this is the only time in the history of life on Earth that land animals this massive have occurred. Ecolology of the ornithischians is perhaps the most easily conceptualised; being large herbivores at about 5 tonnes body weight, possibly analogous to the large herbivores of today, such as rhinos and elephants. Their predators are harder to understand because giant, 5 tonnes carnivores no longer exist, nor did so previously. Perhaps, abundant prey was plentiful enough to support them and one large predator possibly represented the ecological occupation of a group of pack-hunters, which is typically the predatory paradigm today. Most incomprehensible are the enourmous sauropods, which principally exemplify evolution of large size. No terrestrial animals this large have ever existed before or after this era in the history of life on Earth.

Evolution of large size, however, is not exclusive to the dinosaurs, albeit exqiusitely exemplified by them. Early fish in the Devonian, such as *Dunkelosteus* and later *Charcharadon* grew to large sizes, as did some mammals in the Tertiary, such as the 30 tonnes rhinoceros, *Indiricatherium*. So, why evolve large size? An evolutionary principle, known as Cope's rule, states that evolution tends towards large size; "biggest is best", so to speak. The advantage of being big is reduced susceptibility to predation. The large adult sauropods would essentially have been immune to attacks by carnivores. Intraspecifically, the strongest members of a herd are less likely to be targeted by predators than the old, young and sick. Hence, the biggest and strongest herd members are preferred for breeding and are successful in mating, while the weaker competition is excluded, for example, red deer ruts. As a result, populations of animals may grow larger in stature over time. In response, following trend of the perpetual arms race in nature, predators may become larger to tackle the big herbivores, for example the carnosaurs, and later sabre-tooth cats in the Tertiary megafauna. There are also physiological benefits to large size, such as digestion and heat retention. The problems with being very large is that such animals have very specialised ecological niches, there is less scope for adaptation, so there is a larger risk of extinction. Big animals need large amounts of food, so have small population sizes. Evolutionarily speaking, they do not persist through times of adversity, for example cynodonts survived the P-T, pelycosaurs didn't and small mammals surpassed the dinosaurs at the K-T extinction.

Although evolution might tend towards large size, true gigantism is rare because of the ecological implications, and partly due to the mechanical constraints involved. Large bodies are difficult to design and produce. The size of a terrestrial vertebrate is limited to the strength of bone and the power of muscle. Bones must become much more robust, having a larger cross-sectional area (a square value) as animals evolve larger size, given that increase in body volume is a cubic relationship. Bones must have strength in excess of the minimum value needed to support the body's weight, due to the stress of force from body weight being expressed at an angle to the bone during locomotion, causing bending as well as compression, which increases the susceptibility of bone breaking. This is helped by the pillar limbs set for erect posture. The muscle, meanwhile, must be powerful enough to be able to push the body up from a lying postion and move the body in locomotion. Biomechanical calculations estimate that an 80 tonnes *Brachiosaurus* is about at the threshold of size that a terrestrial animal can achieve. It would have been reduced to a sedate walking pace, so as not to overload the bones with stress.

Large size in sauropods may be an extreme of the "big gut solution" to herbivory, while the long neck was an adaptation to reach the tops of trees, like a giraffe. Food was nipped off by peg-like teeth, then macerated by gizzard stones, or gastroliths, before slowly passing through a long labyrinth of the gut to maximise digestion of the low quality forage. Giraffes need not be so heavy because thay are ruminants, and regurgitate a bolus of plant material to chew in the molars during cudding. This makes plant digestion more efficient, without the need for as much fermentation in large, heavy guts. Similarly, ornithischian dinosaurs had batteries of grinding teeth, so digestion in the guts was not as extensive.

The already diverse dinosaurs radiated further in the Cretaceous period 135–65 Ma. The Cretaceous period was a time of abundant volcanic activity as the continents continued to separate, with India and Madagascar parting from the land mass known as Gondwana. This created global warming. Significant floral changes also occurred during the Cretaceous: the seed ferns, conifers and cycads gradually became replaced by angiosperms, or flowering plants. The Mesozoic forests were initially made up of the conifers and cycads. The conifers grew to an average height of about 30 m, having a long, slender trunk that terminated in a canopy of needle-shaped leaves, among which the seed-bearing cones were located. The palm-like cycads also had leafless stems, crowned with a mass of stiff, large leaves at the centre of which protruded a seed-bearing cob. Angiosperms were different, coevolving with insects, having flowers as part of pollination, e.g. magnolia, beech, fig, palm.

During the Cretaceous, the other division of the thyreophora, the ankylosuara, became common. Also, the cerapoda clade of the ornithischians radiated to include new and important groups. The early ankylosaurs, such as *Polarcanthus* from southern England, had spiny plates along their flanks. As the ankylosaurs evolved, later forms, such as *Euplocephalus*, developed more elaborate body armour, formed as plates rather than spines. These covered the neck, trunk and tail. Also, *Euplocephalus*, like many other ankylosaurids, had massive bony bosses at the end of the tail, formed from the fusion of the terminal vertebrae and incorporation of bony skin plates. A 7 m long *Euplocephalus* could swing its tail club, delivering an incredibly powerful blow, which would readily disable an attacking *Tyrannosaurus*.

The skull of ankylosaurids was a heavy box-like structure. Overgrowth of bones in the skull formed thick bony studs, while the bone plates formed within the skin over the head fused with the skull, covering the temporal fenestrae, leaving only two gaps either side of the skull: a small nostril and an orbital, which had a bony eyelid. It is obvious that the ankylosaurids chose body armour, rather than speed, in the long-term evolutionary battle between predators and prey. Like other thyreophorans, ankylosaurids had broad beaks to crop low-lying vegetation, such as ferns and the soft-stemmed flowering plants that were spreading north at this time. Small teeth in the cheek may have sliced up soft plant material, but the food was mainly mashed up in the gut.

The cerapod clade became very diverse in the Cretaceous. This group included the bone-headed pachycephalosaurs, the horned ceratopsians, and the diverse group of ornithopods ancestral to *Heterodontosaurus*. The pachycephalosaurs were characterised by their conspicuous skull morphology; having a remarkably thick skull roof in which the bones of the skull fused in a solid dome, which may be up to 22 cm thick in a skull that is 62 cm long in some genera. Also noteable, was the long shelf jutting out from the back of the skull, which was often ornamented with bony studs. This bony shelf is also seen in the ceratopsians; thus the pachysephalosaurs and the ceratopsians are united in a group called the marginocephalia (literally meaning, "margin-headed ones"). Pachycephalosaurs were bipedal herbivores. Some early ceratopsians shared this trait, although most were quadrupedal. It is speculated that pachycephalosaurs, such as *Stegoceras*, used their thick skulls in butting contests over the right to mate, like modern sheep and goats today. This is perhaps why the skulls of the males are thicker than those of the females. Quarrelling *Stegoceras* would have adopted a horizontal backbone posture when charging so as to dissipate the impact of the force evenly through the skull and straight down the neck to the shoulders and hindlimbs.

The ceratopsians were a much larger group than the pachycephalosaurs, including twenty genera, mostly from the late Cretaceous of North America. Similarly to the stegosaurs, ceratopsians had longer hindlimbs than forelimbs; indicative of a bipedal ancestry. The relatively long limbs, combined with the digitigrade posture, meant that ceratopsians were adapted for galloping, unlike cumbersome ankylosaurids. Early forms, such as *Protoceratops* from east Asia, already have the ceratopsian hallmarks: a deep snout, frill and the begginings of a nose horn. A famous fossil discovery was found in the Gobi Desert in 1971 that depicts a *Protoceratops* in combat with a *Velociraptor*. It may have been that the two died as a result of their injuries, or perhaps they became buried in a sand dune as they fought.

The later ceratopsians were larger than the 2.5 m long *Protoceratops* and had tall neutral spines on the vertebrae of the neck. These are associated with muscle attachment: powerful muscle were required to hold their massive heads up. The skull may make up a third of the animal's entire body length. Indeed,

Torosaurus had a 2.6 m long skull and the frill was larger than the skull itself. The name ceratopsia means "horned faces", which refers to the variety of horns displayed by the ceratopsians. Some, like *Centrosaurus*, had a single nose horn, formed by fused nasal bones. Others had an additional pair of horns on their brow.

Styracoaurus was a centrosaurine, having an elongated nasal horn. It also had six prominent backward-pointing spikes on the edge of the frill, which would have looked imposing when a 5 m long *Styracoaurus* dipped its head in display. The most famous ceratopsian, *Triceratops*, had a relatively short frill that was fully enclosed in bone. Growing up to 9 m in length and weighing about 6 tonnes, *Triceratops* were among the last large herbivorous dinosaurs, appearing in the late Cretaceous 68 million years ago.

Ceratopsians would have lived in herds, feeding on low-growing plants within open woodland. The frills and horns were probably important features of a ceratopsian functioning within a herd. They may have had a role as signalling structures to identify members of the same species as well as threat displays. Male ceratopsians may have partaken in herd wrestling, with the horns interlocked, like deer do today, when competing for a mate. Also, the frills and horns may have been for defensive purposes. The frill, when the head was lifted, would protect the neck, while the horns could readily impale an attacking carnosaur. For this reason, demonstrations of the frill and horns in a sexual display are justified, in that males with the strongest frills and largest horns would be more capable of warding off predators, thus these successful genes were desired by the females.

Living alongside the ceratopsians, more advanced ornithopods became significant in the late Cretaceous. Derived from small bipedal herbivores in the early Triassic, such as *Heterodontosaurus*, larger bipedal herbivores evolved; the likes of *Iguanodon,* and a group collectively known as the hadrosaurs. Fossils of *Iguanodon* were officially discovered in England in 1825; then to be officially named after *Megalosaurus* (a bipedal carnivore), but the nature of *Iguanodon's* identity was controversial. It wasn't until 1878, when numerous complete 8–12 m long skeletons were found in a coal mine in Belgium that

Iguanodon was desscribed accurately. *Iguanodon* had a vaguely horse-like head, with a beak and cropping teeth in the cheeks, which it used to graze low-lying fronds and branches and the leaves of tree ferns and flowering plants, such as magnolia. The hands were highly specialised, with several modifications to the anatomy: fused carpal and metacarpals, forming a block in the wrist; digit one is reduced to a thumb spike; digits two-to-four form a bunch, where digits two and three have small hooves. *Iguanodon* could evidently use its forelimbs for walking as well as being bipedal. The thumb spike was likely used in defence or display.

The hadrosaurs, or "duck-billed" dinosaurs were the most sucessful and diverse ornithopod clade. Hadrosaurs appeared later in the Cretaceous than *Iguanodon*; specimens are known from North America, central Asia and China, where they roamed the lowlands in mixed-species herds. Hadrosaurs were predominantly bipedal, but like *Iguanodon*, hooves on their fingers suggest that they could also walk on all fours. With the skeleton in a horizontal posture, hadrosaurs were well-adapted for efficient running; the mechanism used by this group to avoid predation.

The typical hadrosaur skull has the distinctive duck-like bill, where the premaxilla and the maxilla are flattened and further back. The teeth are adapted for grinding and consist of upto five or six long rows comprised of forty-five or sixty teeth, set well back from the bill. These batteries of teeth were necessary because rapid wear occurred around the jaw margin due to lateral movement of the jaw.

While the general body plan of hadrosaurs is uniform, the variety of hadrosaur genera are readily differentiated by the conspicuous array of headgear displayed by the lambeosaur group. The premaxilla and the nasal bones are extended forming crests, which can be used to distinguish between different genera. Some had high flat-sided "helmets", such as *Corythosaurus*, while others may have forward-directing rods, as with *Lambeosaurus*, others, like *Parasaurolophus*, had backwards-pointing tubes. What were these crests used for? It is known that the nasal cavities extended from the nostrils into the crest as four air passages: two run from the nostrils and two run from the crest to the throat region, so air breathed in through the nostrils had to take this route. There have been several speculative postulations: snorkels, salt glands, olfactory enhancers, but the most favoured explanation is the use of the crests as signalling devices. The shape, size and, perhaps, colour of the crest may have been used by visual species as a means to identify members of a species, recognise potential mates and to signify positions in dominance hierarchies. Males and females appear to have different crests. Furthermore, the shape of the air passages within the crests would have enabled hadrosaurs to propagate a low resonating note as air was inhaled. Since the shape of the air passages varies between males and females and juveniles, the noise they made would be different. This would help identify gender and age between same-species members of the herd. Different species, with different crests, would produce a noise even more markedly different. Herds of hadrosaurs composed of several species would have augmented a conspicuous auditory display in the Cretaceous landscape as they strived to evaluate their companions. The noise may also have been used to deter predators, particularly *Parasaurolophus*, which would have produced an exceptionally loud, blaring deep tone as the herd called out harmoniously. This would surely be off-putting to any nearby predator.

In addition to the abundant herbivores that evolved during the Cretaceous, the carnivorous theropods also diversified greatly. A highly specialised group, the ornithomimids, arose during this period. They had a slender ostrich-like body, with long limbs. The hands had three powerful fingers for grasping prey and the long limbs enabled ornithomimids, such as *Struthiomimus*, to run exceedingly fast: speeds ranging from 35–60 km/hr have been estimated. The skull in later forms is completely toothless and the diet consists of small prey, such as lizards and mammals and, perhaps, even omniverous habits as well. These dinosaurs therefore, occupied different ecological niches to the more fiercesome carnosaurs that targeted larger prey.

One major theropod clade of the Cretaceous were the maniraptora. This group includes earlier forms from the Jurassic, such as *Compsognathus*; the smallest adult dinosaur at 70 cm in length. Of the Cretaceous groups, the dromaeosaurids were the most diverse. This group includes the pack-hunting dinosaurs, such as *Velociraptor* and the larger *Deinonychus*. *Deinonychus*, found in Montana, USA, like other dromaeosaurs was a relatively small dinosaur at 3 m in length, a metre tall and weighing 60–75 kg. The curved, sharp serated teeth, present in all theropods, indicates the carnivorous diet of *Deinonychus*.

As a hunting strategy, *Deinonychus* would operate in a pack, where each member relied on speed and agility to bring down prey animals larger than themselves, and were adapted accordingly. During running, *Deinonychus* adopted a horizontal posture, where the tail and backbone were continuously linear. This allowed the body to be balanced correctly, with the centre of gravity located at the hips, while the neck held the head up in a swan-like S-shape. The tail remained in this horizontal position because it was stiffened by long, bony rods in the middle and posterior sections of the tail. The long arms of *Deinonychus* were srong with large hands, which indeed were almost half the length of the arm. Deep claws on the end of its three long fingers; hands that could turn inwards due to a highly mobile wrist joint, clearly suggest that the hands were used for gripping prey. The long hindlimbs have bird-like proportions: a short femur, long tibia and fibula, long metatarsals and three functioning toes, with a backwards-pointing first toe. The key fearure of the foot is the elongated second toe, armed with an enormous sickle-shaped claw up to 12 cm long. Indeed, the name *Deinonychus* means, "terrible claw". It would have been used to slash at prey and disembowel them. While walking, the claw was held upright to avoid getting in the way. By hunting in packs, *Deinonyvhus* could tackle larger prey, such as 6–7 m long hadrosaurs.

The most famous and notorious dinosaur ever to have lived; synonynous among the big meat-eating theropods, arose at the very end of the Cretaceous. *Tyrannosaurus rex* was among the largest terrestrial predators ever to have lived. The tyrannosaurids (tyrant lizards) represented the pinnacle of carnosaur evolution, radiating in North America and central Asia. *Tyrannosaurus* was well-adapted for hunting the larger Cretaceous herbivores, namely hadrosaurs and ceratopsians; having remarkable size and power and immense jaws. A T.rex could grow as large as 12–14 m in length and weigh 6–7 tonnes, with the females thought to be larger than the males.

The evolved characteristics of *Tyrannosaurus* were focused around the skull. The forelimbs were absurdly small, perhaps to help balance the 1.2 m long head, which was supported by thick neck muscles. As a consequence, the forelimbs were, apparently, useless; not even being able to reach the mouth. Although, some scientists believe that the small arms played a role in helping *Tyrannosaurus* to get up from a lying position, providing a push while the head is thrown back and the legs straightened. The jaws contained over fifty dagger-like teeth at 15 cm in length. Additional holes in the skull are present, presumably to alleviate the great weight. In addition to the temporal and antorbital fenestrae, carnosaurs also had a maxillary and mandibular fenestra. These may also have acted as sites for muscle attachment, provding T.rex with its irrefutably powerful bite, which was wide due to the additional joint in the lower jaw, which increased gape.

Tyrannosaurus was well-adapted to tackle large prey herbivores, but would also opportunistically prey on the old and sick, or scavenge. There is some controversy over the presumed hunting methods of T.rex. Some scientits believe that T.rex would have ambushed prey; attacking at a speedy walk, while others think that it was capable of running at speed on its three toes in a digitigrade stance, and run prey down. What is certain, *Tyrannosaurus*, which lived 68–65 million years ago, was the most advanced large carnivorous dinosaur; the product of 165 million years of dinosaur evolution.

Dinosaur fossils have been found in Cretaceous rocks of Alaska and Australia, along with deciduous plants that shed leaves; not needed during long, cold, dark winters. These continents were at 70° N and 80° S, respectively. At these polar latitudes, dinosaurs would have been subjected to seasonal changes of light, temperature and food supply; conditions that don't suit modern reptiles. Herds of hadrosaurs would have migrated when conditions became too cold and food scarce, followed by tyrannosaurids in pursuit, like how wolves follow migrating caribou today. Being reptiles, dinosaurs were presumed to be ectotherms, but this is an assertion that has been under much scrutiny. Some palaeontologists believe that dinosaurs may have, in fact, been warm-blooded, but the nature of dinosaurian thermoregulatory physiology is still subject to much scepticism. Rather than relying on heat from the external environment to warm their bodies, certain lines of evidence have been interpreted as dinosaurs being able to generate body heat internally through metabolism, which would certainly have helped the polar-dwelling communities. This evidence will be outlined briefly here.

Dinosaurs have an erect stance and more advanced gait than their predescessors, similar to endotherms (mammals and birds). This posture enabled some of them to run at high speeds, which would require metabolic energy. However, this only applies to small bipedal dinosaurs; the larger ones will have likely locomoted in a lumbering walk. Also, there is no demonstrable cause between endothermy and erect gait. Small dinosaurs have high activity, but snakes and lizards can move fast in short bursts.

The haemodynamics of sauropods considers the problems of pumping blood up the long neck to the brain. To do this, it is postulated that sauropods must have had a strong four-chambered heart, like endotherms today. This correlation isn't certain because crocodilians also have four-chambered hearts. Bone histology research on dinosaurs has shown that dinosaurs had highly vascularised bone, quite like that of mammals and unlike those of many living reptiles, similar to haversian bone systems. That said, some reptiles today have haversian bone, despite being ectotherms and some highly active endotherms have no haversian systems. Furthermore, many dinosaurs had fibrolamellar bone, which occurs through rapid growth, without the formation of growth rings, and is present in the large mammals today. Fibrolamella bone only implies fast growth rates, but dinosaurs often show lamellar-zonal bone indicative of episodic rapid growth rates, which, perhaps, suggests a mixed thermoregulatory regime.

Brain size is another factor: while most dinosaurs had small lizard-like brains, ornithomimids, dromaeosaurs and troodontids had large bird-like brains and active brains probably require endothermically derived energy. Indeed, endothermic birds are derived from these theropod dinosaurs, so perhaps among these clades there is an observed transition to endothernic bird-like physiology; moving away from ancestral reptilian affinities. However, the large brains of these dinosaurs are associated with

101

good eyesight and balance, and don't imply advanced mammal-like intelligence.

Many scientists now believe that larger dinosaurs were neither ectothermic nor endothermic; instead they had an intermediate condition: inertial homeothermy. Experiments on large living reptiles show that rates of internal temperature change are very slow in normal subtropical conditions. Large dinosaurs have a very small surface area-to-volume ratio, so would lose very little heat, and therefore would remain in homeostasis inertially, without having to produce internal body heat. They would have kept warm through activity metabolism rather than thermoregulation. Digestion of food in the large guts would also have produced a lot of internal heat; another advantage of large size. However, small dinosaurs, like hypsilophodontids, were active and possibly endotherms. Through the 165 million years of evolution through the dinosaur age, these dominant reptiles of the Mesozoic became extrordinarily diverse; occupied many ecological niches and climatic settings and displayed many adaptations to do this, including large size and endothermy. The dinosaurs weren't the only reptiles of the Mesozoic, of course. One environment that the dinosaurs didn't diverge towards was marine habitats. The Mesozoic sea serpents were a discrete clade of their own.

Mesozoic Seas

Since evolving from their amphibian ancestors, the reptiles evolved and diversified greatly. They survived the great mass extinction at the end of the Permian and thrived through the Mesozoic Era, becoming the dominant land vertebrates. The reptiles were so successful because, at the time, they were the most well-adapted class of vertebrate animals for life on land. Impermeable skin prevented them from dessicating, pillar-like limbs made efficient locomotion and advanced physiology enabled them to breathe air effectively and feed on land, becoming totally independent of the water. However, in the Mesozoic, some reptiles took to the water and became adapted to an aquatic mode of life.

The marine reptiles were differentiated from their land cousins in possessing the euryapsid skull, with one (upper) temporal fenestra. All the main groups originated in the Triassic and have very different adaptations. These groups are the nothosaurs, placodonts, ichthyosaurs and the pleisiosauria. One feature these reptiles have in common is that they are all predators, but the variation in adaptations suggest that they were derived from independent origins.

The nothosaurs were elongate animals, varying in length from 0.2–4 m, with long necks and tails and a small skull. The paddle-like limbs were attached to reduced limb girdles, so could have supported the body out of water, but likely functioned to steer the body as the animal propelled itself through the water using its deep tail. The limbs would have been held close to the body to reduce drag as an agile nothosaur, such as *Pachypleurosaurus*, pursued the fish that it preyed upon. It is possible that nothosaurs may have evacuated the ocean for respite on the beach when marine predators were nearby.

The placodonts were also abundant during the mid-Triassic, but died out towards the end of the period. Upon first sight, they appear to be land adapted, being heavily built and the limbs lack any marine adaptation. However, the limb girdles are not robust enough to support the body out of water; it is likely that they lived in shallow marine environments and walked along the bottom to search for invertebrate prey. The teeth are covered with heavy enamel for crushing the hard shells of molluscs, once they were prized from the rock.

The ichthyosaurs (meaning fish lizards) had much more obvious aquatic adaptations: no neck, a streamlined body, paddles and a fish-like tail. These reptiles were more suited to open water: a niche which they successfully exploited since they endured throughout the Mesozoic Era, with little change to the body plan. Some late Triassic ichthyosaurs reached lengths of 15 m, but in the late Jurassic and Cretaceous forms never reached this size again, but common features such as a long snout and large eyes were retained. Fossils of the stomach contents show hooklets from squid tantacles and fish scales, which elucidate the animal's diet. Invertebrates were less common as a food source because the ichthyosaurs couldn't access the soft bodies of ammonites and belemnites from the hard shells. Ichthyosurs were totally marine reptiles: unlike turtles and pleisiosaurs, they could not return to the land to give birth. Remarkable fossil evidence from the early Jurassic of Germany shows how live young were born tail first; the mother must have died giving birth, struggling for air during the unfortunate event.

The first pleisiosaurs appeared along side the ichthyosaurs in the late Triassic and are believed to be closely related to the nothosaurs, albiet larger, ranging from 2–14 m in length. All had powerful paddle-like limbs, well adapted for submarine locomotion. The body morphology of pleisiosaurs varied though: some, like the criptoclidids, had long necks and interlocking pointed teeth for capturing and holding slippery fish, while others had relatively short necks and a long, heavy skull, such as the pliosaurs. Massive pliosaurs, such as the 12 m long *Liopleurodon*, may have preyed on smaller pleisiosaurs and ichthyosaurs.

The mechanism of locomotion of plesiosaurs is a topic of speculative intrigue. The large and powerful paddles are presumed to be the source of thrust, since the tail is relatively small, but the mechanism used to coordinate the paddles and generate propulsion is under much debate. Originally, it was thought that plesiosaurs "rowed" by beating the paddles backwards and forwards, but the back stroke would have created a counter thrust that would cancel out forward motion to some extent and the paddles could not be feathered (i.e tipped to the horizontal). A more favoured hypothesis demonstrates the underwater "flying" mechanism, used by tutles and penguins today. The flat paddle, which has an aerofoil crossection, when pushed down and backwards at the right angle above the horizontal; lift is generated with forward propulsion. The tip of the paddle demonstrates a figure of eight, where each part of the cycle produces forward movement. A modified version of this hypothesis, however, describes the paddles being moved in a cresent-shaped path, like a sealion. Analysis of the anatomy of plesiosaurs reveals that the heavy limb girdles would have restricted the upward movement of the paddles. Using the revised latter mechanism, pleiosaurs could attain quite fast speeds, particularly pliosaurs, making them effective marine hunters.

Evolution of Modern Reptiles

Besides the dominant and spectacular dinosaurs, other groups of modern reptiles evolved during the Mesozoic. Many of the Mesozoic reptiles were derived from Permian-age forms. The dinosaurs, and closely related pterosaurs, evolved from the thecodontian group of archosaurs, forming part of an ornithosuchan lineage, along with the Triassic crocodilomorphs. The four major groups of marine reptiles; nothosaurs, placodonts, ichthyosaurs and pleisiosaurs are all thought to be derived from diapsid archosaurs also. The archosaur branch diverged from the lepidosauromorphs in the Permian, which came to include the lizards and snakes. The turtles and tortoises (testudines) form a separate clade from the diapsids that predates the Mesozoic. From the late Cretaceous onwards, significant radiations of modern reptiles occurred.

Global warming in the Cretaceous, attributed to the volcanism as India and Madacasgar seperated from Gondwana, brought about floral changes. As seed ferns, cycads and conifers became replaced by flowering plants, the angiosperms, due to the changing climate, new ecological settings were produced. The evolution of flowering plants made habitats that were more akin to modern ecosystems than previously. During this time, reptiles bearing more resemblance to modern faunas developed; being adapted to these more familiar niches. Mesozoic gynosperms were the dominant land plants in the Triassic and Jurassic and were the staple food of herbivorous dinosaurs and other reptiles. By the Cretaceous, flowering angiosperms quickly became dominant and offered a new source of food and habitat to many animals, including modern reptiles.

Seed plant reproduction has two main phases: fertilisation, where the plant must be pollinated, and seed dispersal, where the protected plant embryo is transported to a favourable location to germinate. Conifers, and many other plants, are pollinated by wind; an enormous number of pollen grains are released to blow in the wind in the hope that a pollen grain will reach the pollen receptor of a female plant organ of the same species, for example in the conifer cones. To maximise the chance of pollination, the pollen is released in dry weather, with optimal wind conditions and female cones are aerodynamically shaped to act as effective pollen collectors and become saturated with pollen. However, the process is still expensive; producing many gametes, each with a small likelihood of successful fertilisation. Consequently, these species tend to exist in homogenous communities, such as temperate forests. Wind pollination doesn't work in mixed species communities, like a modern jungle.

The angiosperms made a major step in plant evolution and used flowers to lure animals, predominantly insects, to help fertilisation and disperse the seeds. It is possible to envisage ungainly beetles, which were common during the Jurassic, feeding on the stockpiles of pollen ready to be released when conditions became suitable. These beetles would then move to another plant, maybe of the same species as before, covered in the pollen of the previous plant and, therefore, act as an effective vector to transport the pollen. Over time, the plant structures may have evolved to cooperate with insects, such as making pollen more available, while protecting delicate parts, and producing alluring scents and colourful flowers to entice them towards a nectar-rich reward. In this way, insects deliver pollen directly between plants; much more effective than chance pollination by wind. Insects meanwhile coevolve, becoming loyal to the plants; those insects that learn the particular scent and colour of a flower evolve rapid and error-free recognition of pollen sources and become the best food gatherers and succussful reproducers. In fact, today insects discriminate strongly between these features and may also congregate for mating around certain plant species. The evolution of the angiosperms correlates strongly with the increased diversity of insects in the Mesozoic. By extension, more food sources were available to animals that fed on them, such as modern tree-dwelling lizards as well as the first birds. Modern reptiles also adapted to feed directly on the nutritious nectar of angiosperms. The evolution of these plants in the Cretaceous was, therefore, an important change in ecology, which set the path towards the evolution of modern faunas.

Crocodilians are very much a noteworthy example of modern reptiles; being evocative of the prehistoric reptilian beasts. Although they are superbly adapted to their habitat today, they, in fact, originate from before the dinosaurs and have modified their persona over time to cope with the environment within which they were living. Today, crocodilians may be found in both fresh and salt water of the tropics and constitute eight genera of crocodiles, alligators and gavials, which typically prey on mammals along the water's edge and display habits accordingly. On land, crocodiles exhibit four modes of locomotion: belly run, where the body is pushed by the hindlimbs for escape down riverbanks; sprawling, for slow locomotion; high walk, where the limbs are tucked under the body for faster movement; and finally, galloping, where the forelimbs and hindlimbs unexpectedly act in pairs. Typically though, they don't pursue their prey directly on land by chasing it down. Instead, their technique is to lunge from a concealed position in the shallow water's edge and grab any drinking animal in its powerful jaws.

This is not how crocodilians have always existed. The first crocodilomorphs from the Permian were lightly built and probably bipedal. The first true crocodilians from the early parts of the Mesozoic, were small (about a metre long) and quadrupedal, such as *Protosuchus* and *Ornithosuchus*, whose hindlimbs were longer than the forelimbs, portraying their bipedal ancestry. Around 150 genera of fossil crocodilians from the Jurassic and Cretaceous have been classified in a group called the mesosuchia, which are mainly aquatic forms that lack the specialisation of living groups, the eusuchia. *Geosaurus*, from the late Jurassic of Europe, was adapted for an entirely aquatic existance to the extent that it would have had difficulty walking on land. Powerful undulations of the body generated propulsion from a tail fin; the limbs are paddle-like, and body armour is omitted to improve hydrodynamic efficiency. The true crocodiles, or eusuchians, appeared in the late Cretaceous and were more abundant than today.

The lepidosauria, which became distinguised in the Permain, includes the sphenodontians and squamates: the lizards and snakes, respecitvely. These groups became very diverse in the Mesozoic, with a variety of diets; insectivory, herbivory and, perhaps, some aquatic forms. The lizards radiated into six main lineages in the late Jurassic and Cretaceous. The gekkotans, which includes the tiny geckos that can cling to walls and ceilings, the iguania, represented today by iguanas and chameleons, the skincomorphas, including today's skinks, the anguimorphs, including living monitor lizards, the amphisbaenians, adapted for a burrowing lifestyle, and, finally, the snakes, which are believed to have arisen from lizard ancestors. Snakes first appeared in the early Cretaceous and radiated greatly onwards into the Tertiary along with the mammals on which they preyed. Originally, snakes killed their prey by coiling and causing aysphyxiation, as boas and pythons do today, prior to the later forms evolving venom.

The order testudines, the turtles and tortoises arose in the late Triassic; they are differentiated from most other reptiles by having no temporal fenestra; they are anapsid and later forms lacked teeth. Testudines diversified greatly due to a successful accessory: the shell. A protective capsule into which the animal could retreat in the presence of a threat meant that only predators that were strong enough to break the shell could kill them. Particularly for some of the prehistoric turtles, which grew very large, this significantly limited the number of potential predators. The shell is made in two parts: a domed carapace on top and a flat plastron below, which they made from bony plates growing within the skin, attached to the vertebrae, ribs and shoulder girdle. Modern turtles, from the Jurassic to present day, are classified into two main groups: the pleurodira and the cryptodira, defined by the way the animals retract their heads into the shell in the presence of danger. The pleurodires pull the head in by making a sideways bend in the neck, while cryptodires make a vertical bend.

By the Cretaceous, modern reptiles became a significant aspect of the Mesozoic faunas. While the tetudines had success in the water, on land the evolution of angiosperms created habitat and food for small lizards; competition from their dinosaur relatives was minimal here, so modern faunas were able to fluorish. Crocodilians, being larger animals, would have been in direct competition with some kinds of dinosaurs, but adaptive supremacy meant that they were able to coexist, but also endure beyond the Mesozoic along with their smaller lizard cousins.

Pterosaurs, Birds and Flight

Throughout the Mesozoic, dinosaurs and their more familiar reptilian counterparts dominated the land. Reptiles also diversified to exploit the ecological niches available in the water:

the marine euryaspids, crocodilians and testudines were all well-adapted for an aquatic lifestyle. Remarkably, reptiles even evolved the ability to fly: the pterosaurs exploited the hitherto vacant niche, which provided them with abundant opportunities. Reptiles became the first vertebrates to conquer the land, sea and air.

The name pterosaur means, "winged reptile" and they existed throughout the Mesozoic alongside their close relatives, the dinosaurs. The characteristic body shape of the pterosaur was developed from the offset, when the first pterosaurs, such as *Eudimorphodon*, evolved in the late Triassic. They had a short body, fused hip bones, a long neck with large head and pointed jaws; most importantly, though, the specially evolved arms for flight. The joints in the arm and wrist were involved in folding the wing when grounded and the hand had three grasping fingers and an elongate fourth finger that supported the wing membrane. The two wings extend from the tip of the long fourth finger down towards the animal's hindlegs; their kite-like shape would be silhouetted against the sky, with the long neck protruding infront. To this fundamental body plan, different genera exhibited idiosyncrasies pertaining to their specific ecological role.

Diversification of the pterosaurs occurred notably in the Jurassic and Cretaceous and is most apparent when studying a selection of skulls. Most obvious is the size: while the 90 mm skull of *Eudimorphodon* is little larger than that of a pigeon, the skull of a *Pterodon* may be 1.8 m long; a much larger animal. In regards to anatomical observations, notable evolutionary changes also occurred in the skull: the forward shift of the jaw point to lie below the orbit, elongation of the skull and fusion of the nostril with the antorbital fenestra. As different genera of pterosaur specialised to their individual niches, the variety of food sources they directed towards can be interpreted from the teeth. *Rhamphorynchus* had long, spaced teeth for gripping fish, which it caught by trawling the lower jaw through the surface sea water. By contrast, the short teeth of *Dimorphodon* were probably used for insect eating. More uniquely specialised are the 400–500 flexible teeth of *Pterodaustro*, used to strain microscopic plankton.

One of the largest pterosaurs, with a wingspan of 5–8 m, *Pterodon* had few teeth and probably swallowed the fish immediately after extraction from the sea. The skull is longer than the trunk, and half its length is due to the pointed crest at the back, which may have acted like a weathervane to keep the head pointing forwards in flight. *Pterodon* was built sturdily: each of the massive cervical vertebrae have a pneumatic foramen; a hollow containing an air pocket to alleviate weight without compromising on strength, forming a light-weight but robust skeleton. The dorsal vertebrae are involved in stabilising and supporting the shoulder girdle and pelvis. The shoulder girdle is attached to a sturdy sternum, which has a slight keel shape for the attachment of large flight muscles. The skeletal rigidity is likely to be due to the stresses of landing impact, rather than flight stresses. When in flight, the skeleton would indeed have to cope with the stress produced by strong flight muscles in operation, but this tension would be incomparable with the trauma of such a large animal landing clumsily on the ground. Albeit very lage, *Pterodon* was not the largest pterosaur: known from part of a single wing, *Quetzalocoatlus* from the late Cretaceous of Texas had a wingspan of 11–15 m, similar to the size of a small aeroplane; the largest flying animal of all time. This creature would have looked magnificent flying at a low altitude as it swooped onto land; thudding the ground as it completed its careful descent.

The most obvious specialised adaptation of the pterosaurs is the ability to fly. In the past, pterosaurs have been portrayed as inefficient gliding animals and, on the ground, clumsy, staggering bat-like movers. However, recent views consider the pterosaurs as been well-adapted for true flight. Firstly, pterosaurs have a number of flight and aerodynamic adaptations, such as moving wings, hollow bones and reinforced landing gear and a streamlined head. The second line of evidence suggests that pterosaurs were probably endotherms, having hair, shown in some well-preserved specimens; only endotherms have insulation, and endothermy provided pterosaurs with the sustained metabolic rate necessary for flight.

The wing is composed of skin, forming a membrane attached to the side of the body and extends to the tip of the elongated flight finger, four. The wing membrane is reinforced by stiff elastic fibres that run parallel to one another. Well preserved specimens show that the pterosaur wing was slender, like that of a gull, rather than a broad gliding structure; although in *Pterdactylus*, at least, the membrane extends to the femur. To attain lift and take-off, the pterosaur directed a power stroke down and forward, while the recovery stroke moved up and backwards, so that the tip of the wing moved in a figure-of-eight. Pterosaurs probably took-off from trees or cliffs after a short run-up. Generating momentum off the ground would be difficult, particularly for larger genera, so would require a larger run-up, or indeed be impossible, putting the animal at risk of predators.

The evolution of flight was a remarkable acheivment and provided pterosaurs with an abundance of opportunities. They diversified to pursue different food sources by implementing novel hunting strategies, thus avoiding competition with other vertebrates. Initially, their prey would have been poorly adapted to evade aerial predation, so pterosaurs thrived. Also, flight was a superb way to avoid their own predators: a quick escape to the sky and they were out of the reach of land predators. They could seek refuge at the top of trees and cliffs and became very successful.

Of course, pterosaurs were not the only vertebrates to evolve the ability to fly; the first birds evolved in the Mesozoic and very soon, pterosaurs were faced with competition. Birds (class aves) radiated from theropod dinosaurs. Some dromaeosaur dinosaurs appear to have had feathers and, therefore, represent a transition stage from dinosauria to aves. Dromaeosaurs, such a *Velociraptor,* already had bird-like characteristics besides feathers that intimate avian affinities. Long, lightly-built skulls, long arms, which in some incidences may be covered in lengthy plumage, and endothermy. These dinoausrs eventually evolved into genera that closely resembled birds and ultimately, true birds, albeit primitive in appearance; lacking evolutionary specialisations taken for granted when observing birds today.

The oldest known bird is *Archaeopteryx* from the late Jurassic limesones of Solnhofen, Bavaria. It is known from several well-preserved specimens, so its anatomy can be copiously analysed to elucidate the link between reptiles and true birds; for *Archaeopteryx* represents a transient stage in the genesis of hitherto a novel class of vertebrate life. The Solnhofen Limestone formed in a subtropical lagoon. Besides *Archaeopteryx*, it contains specimens of plankton, jellyfish, ammonites, crustaceans, fish, pterosaurs and rare dinosaurs, which sank to the bottom and remained undisturbed. This assemblage illustrates the ecosystem in which the first birds originated.

Archaeopteryx was about the size of a magpie: 30–50 cm long and probably stood about 25 cm tall. It had a lightly-built skull and a bird-like brain, so sight was a key sensory system. Although *Arachaeopteryx* is generally recognised as the oldest known bird, its classification continues to be under debate, since it possesses both bird-like and reptile-like characteristics. *Archaeopteryx* had teeth, separate fingers with claws in the hand at the end of the arm-like structure that constituted a remedial wing. It lacked an ossified sternum and had a long, bony tail. Bird characteristics include a furcula, the fused clavicals (wishbone) and, most obviously, feathers. Indeed, feathers provided *Archaeopteryx* with the means to fly. This was a novel solution to the flight problem; previously pterosaurs implemented a skin membrane, rather than a feathered wing. Despite lacking an ossified, keeled sternum for the attachment of flight muscles, as seen in modern birds, *Archaeopteryx* is considered to be a good flyer. The pectoralis muscle could be attched to the robust furcula and the feathers were asymmetrical and curved, just like in modern birds, enabling the feathers to adjust aerodynamically to all stages of the wing beat.

Archaeopteryx, being the oldest known bird and a competent flyer, meant that there are no fossil representatives of early flight evolution; however, there are a couple of theories. These are: leaping-from-the-ground and gliding-from-trees. The leaping-from-the-ground hypothesis is developed from the idea of small running theropod dinosaurs using feathered arms and tails to gain extra elevation when jumping up to catch flying insects. The gliding-down hypothesis is based on the idea that feathered reptiles were arboreal and could climb trees with their clawed hands and would jump and glide down between trees. By flapping the wings, they could glide further until powered flight was evolved.

End of the Dinosaurs

The Mesozoic lasted for 180 million years, comprising almost a third of the Phanerozoic aeon; throughout which reptiles, particularly dinosaurs, were the dominant land animals and conquered the sea and air too. The Mesozoic started following a great mass extinction and ended 65 million years ago in the same way; bringing about the end of the dinosaur age. Besides the dinosaurs, many other groups were effected in what is known as the K/T extinction. On land, pterosaurs became extinct along with some groups of mammals and birds. Generally, smaller animals survived, along with some crocodilians. Of the marine faunas, free-swimming forms, such as various plankton, the ammonites and belemnites died out; some groups of fish and all the marine reptiles also. Perhaps, most significant was the end of the dinosaurs. Alas, arguably the most majestic land animals ever to have walked the Earth, ceased to exist and life at such large scale would never be repeated.

A number of theories have been postulated to explain the cause of the K/T extinction; the argument is comprised of both catastrophist and gradualist perspectives. The gradualists view of the end of the dinosaurs encompasses a wider time span in which the deterioration of dinosaur faunas occurred, perhaps thousands or tens of thousands of years. The long term factors that led to the demise of the dinosaurs stem from physiological and ecological influences; the dinosaurs, by the end of their successful reign, simply struggled to adapt to the changing environment.

Towards the end of the Cretaceous, the world in which the dinosaurs thrived began to undergo alterations, where the biotic factors of ecosystems changed. Climate tended towards more temperate conifer forests, rather than the subtropical habitat of the dinosaurs. This coincided with marine regression, which effected warm shallow seas as well as terrestrial climate. This produced conditions that were more favourable for the recently arisen mammal communites and the gradualists view is based largely on the competition between the mammals and the dinosaurs, through which the mammals ultimately prevailed.

The evidence to support the gradualists explanation is mostly palaeontological and stratiographic. The Protungulatum mammal community of Montana is from 10,000–40,000 years before the K/T boundary and is devoid of dinosaur fossils, but not so in surrounding sediments. The mammalian populations appear to have spread in correlation with the shifting climate towards more temperate conditions. Around the same time, during the last three million years of the Cretaceous, dinosaur populations declined as many groups became extinct. At this time, only eight dinosaur families remained; represented by only twelve species, such as the theropods *Tyrannosaurus, Albertosaurus*; ornithopod *Edmntosaurus*, ceratopsian *Triceratops*, pachycephalosaur, *Pachycephalosaurus*; and the ankylosaurs, *Ankylosaurus* and *Edmontonia*.

It seems apparent that the mammals were better evolved to adapt to the changing environment. One postulation is that, over 165 million years of evolution, the dinosaurs became subject to a racial senility; the gene pool was "dried up". Holistically, the dinosaurs were such a diverse group, exhibiting gigantism (if not acromegally), exuberant body armour, size and shape variations that eventually their genetic potential became exhausted and were no longer able to adapt readily enough to a changing environment. The mammals, by contrast, represented a new source of genetic variety and readily settled into the new niches; out-competing the dinosaurs.

A catastrophists perception of the K/T boundary depicts a much more abrupt end to the dinosaurs. Extensive volcanism occurred in India, the Deccan Traps, which produced large volumes of the greenhouse gas carbon dioxide, which would have had a profound effect on global climate and caused localised extinction. However, the end of the Cretaceous is marked by an even more abrupt and cataclysmic phenomena: a meteorite impact, the scale of which was enough to finish off the dinosaur faunas.

Topological analysis of the Yucatán Peninsula, Mexico, shows evidence of a crater near a town called Chicxulub, and dates back 65 million years. It is estimated that the crater would have been produced by an asteroid six miles across, and upon impact would have annihilated the local area and caused widespread disaster. The force of the impact caused localised death over a wide area, obliterating ecosystems and would have thrown up dust; blocking out the sun and preventing photosynthesis, upsetting ecosystems far and wide, and caused tsunamis. Further geochemical evidence comes from a world iridium anomaly at the K/T boundary. Iridium enters the Earth from space in meteorites at a low average rate of accretion. At the K/T boundary however, there is an iridium spike, where there is an abundance, suggesting a very large impact. Also, around the presumed site of impact, there is a large amount of shocked quartz, due to the pressure of impact.

The end of the dinosaurs cannot be attributed to either a long-term decline or an abrupt annihilation, but a combination of factors, both long term and sudden, ultimately resulted in the extinction of arguably the most impressive group of land vertebrates in the history of life on Earth. Conditions for the dinosaurs became poor: climate changed, favouring the highly competitive mammals, and the dinosaurs simply struggled to adapt. At the climax of the Creataceous, conditions became increasingly hostile in areas due to volcanism from the Deccan Traps, large animals, particularly the dinosaurs, struggled to survive. Then, at the very end, the asteroid impact wiped out the already dwindling species and afterwards, only small animals could proliferate.

The Rise of Mammals
The First Mammals

Just as the dinosaurs radiated opportunistically after the Permian-Triassic extinction, so did the mammals; proliferating after the Cretaceous event to fill the available ecological niches, following the end of the dinosaurs. With more space available, mammals experienced less competition among themselves and were free from many of the major threats from predation presented to them by the dinosaurs. As a result, the mammals became very successful; their endothermy made them suitable to the changing climate and they were largely insectivorous, meaning they had plentiful food. Consequently, the early radiation of the mammals after the K/T extinction set this class of vertebrates on the path to becoming the dominant land animals throughout the Cainozoic, from 66 million years ago to present day.

The history of the first mammals extends well before the K/T, of course. The first mammalian character arose in the cynodonts of the late Permian, which were derived from synapsid reptiles. Certain groups of cynodonts show a transition stage in the jaw mechanics from a reptilian to a modern mammal-like jaw action. Changes in the jaw, specifically the rearrangement of muscles, led to a profound evolutionary change in the way that cynodonts could chew: unlike their reptilian ancestors, cynodonts could move their jaws backwards and forwards, but also side-to-side, which enabled complex grinding activities. Another change in the jaw was a reduction in the number of cycles of teeth replacement, which was necessary as the teeth became more complex. These modifications to jaw anatomy meant that cynodonts could chew and process their food more thoroughly in the mouth; therefore, the food is utilised more efficiently in digestion.

The general anatomy of the skeleton in cynodonts also showed changes to a more mammalian state. The vertebrae in the backbone became divided into two sets: the thirteen thoracic vertebrae and the seven lumbar vertebrae, which lacked ribs. The tail was also shorter. Changes to the limbs and limb girdles, also occurred; the cynodonts exhibited an erect posture, rather than a sprawling posture like other synapsid reptiles. They had the limbs tucked under the body, so the efficiency of running was improved, since the limbs no longer had to support the body, and stride length could be increased. This was invaluable for them: to be able to sprint away from speedy bipedal dinosaur predators ensured their preservation through the Mesozoic; a time of intense competition for the early mammals under the oppression of the dominant group of vertebrates during the dinosaur age.

The first true mammals appeared at the end of the Triassic. The best known is *Morganucodon* from Europe and China. *Morganucodon* possessed the characteristics of a true mammal, albeit primitive, which showed it to be advanced over the cynodonts. It still possessed the reptilian lower jaw joint, but the bones functioned largely for auditory purposes in the ear. The main jaw hinge was definitively mammalian. The teeth, too, were advanced: cheek teeth may be divided into premolars and molars, while incisors, at the front, shear the food. For small shrew-looking mammals like *Morganucodon*, advances in the teeth were very important. Being able to process and utilise food to the optimum level was vital to efficiently fuel their fast metabolism that was essential for enabling spontaneous bursts of high activity, say, fleeing from a hungry dinosaur, and also to keep a small body warm during cold nights.

As mammals evolved throughout the Mesozoic, their reproductive methods changed; fundamentally important adaptations in their procreative prowess are likely to have been key factors that prompted their success in the Mesozoic and began their path to dominance in the Tertiary. Modern mammals can be divided into three groups, based on reproductive strategy: monotremes, marsupials and placentals; all of which represent a reproductive continuum. The monotremes lay eggs, the marsupials give birth to live young, which, in both cases, finish developing in a pouch, and the placentals retain their young in the uterus to a more advanced stage of development.

Monotremes represent a transient stage in mammalian reproduction, where, although egg-laying is very reptile-like, the parental care of the hatchlings directs towards typically mammalian habits. The pelvis became smaller to improve upon agility and speed in mammalian ancestors, like cynodonts, so eggs had to become smaller, with less food supply in the yolks; hence hatchlings produced were underdeveloped and helpless, so had to be looked after by the parents in safe burrows. However, producing smaller eggs and having rapidly growing hatchlings meant that the first mammals could have rapid reproduction rates, with short intermittencies between litters, which became a crucial evolutionary advantage.

Living monotremes, like the duck-billed platypus, still have a reproductive system presumably comparative to advanced cynodonts and early mammals like *Morganucodon*. Small eggs laid in a burrow produce helpless hatchlings, which have to be cared for and supplied with food. One crucial stage in mammalian evolution was the development of mammary glands, so that mothers could nourish the young by suckling, rather than collecting suitable food for them, enabling the devoted mother to remain with, and protect the hatchlings and offer warmth. Charles Darwin himself speculated on how suckling may have evolved. Mammalian ancestors that incubated the eggs may have had glands that secreted water to keep the eggs humid. Hatchlings that licked these gland would have benefited from the hydration and the secretions may have been antibacterial. Eventually, the secretions contained mineral salts, trace elements and nutritious organic compounds; producing milk meant that excursions for food were less frequent and the young benefited from the increased attendance. Ultimately, specialised nipples developed to supply milk during lactation to young with the necessary cheek muscle adaptations for proper suckling.

The next step in mammalian reproduction was live birth. Live birth became the preferred solution in mammals because a flexible foetus could more easily be passed through a birth canal than a rigid egg and the extra space taken up by a yolk was omitted. Not only was birthing safer, growing the foetus inside the warm body is done at a consistent temperature, optimum for enzyme activity and with an unlimited supply of nourishment. This strategy was only a success because mammals had already evolved suckling, so the mothers were able to tend to the helpless young, which would, otherwise, not survive. In this way, the marsupials evolved.

The next stage in mammalian reproduction was to advance the reproductive organs to transport gas and nutrients more efficiently to the young across a membrane, which was the beginning of the placenta. Keeping the foetus in the specialised sack meant that it could be supplied with nutrients for longer as it grew. Increasing gestation periods meant that the more developed young was less helpless at birth, so the risk of predation was less and parental care was shorter. After a short infancy, young placental mammals would be strong enough to run from predators and the workload of the parents was less, so reproduction could happen more frequently.

Having made advances in their food processing and digestive capabilities, the first mammals, with their functionalised teeth and complicated jaw anatomy, could utilise their food more efficiently. This was imperative for the warm-blooded creatures, with high metabolic rates. It enabled them to live high-energy, nocturnal lifestyles hunting invertebrates in the night; dashing down a burrow in the presence of danger to shelter from the day-time threats. Instrumental in their future success was the advances in reproduction. Particularly with the development of live birth, mammals could reproduce much more often and replenish populations in periods of hardship. This feature alone may be the pivotal reason behind the Tertiary takeover, when the rise of mammals after the K/T extinction enabled them to rise to dominance over the other vertebrate classes.

Tertiary Takeover: Rise of the Mammals

Mammals began their path to dominance at the beginning of the Tertiary period, during the Eocene Epoch, immediately after the K/T extinction. With the dinosaurs gone, mammals were able to thrive. Some of the best-preserved mammal fossils are in the Messel Oil Shales, near Frankfurt, Germany, which document this early takeover. Germany in the Eocene had a humid tropical or subtropical climate; forests of oak, beech, citrus fruits, vines and palms surrounded ponds covered with waterlilies. In these ponds, lived an abundance of aquatic life: fish, frogs, toads, salamanders, crocodilians, tortoises and some lizards and snakes; as well as numerous invertebrates. The preservation of mammals, too, was exceptional, where hair, stomach contents and even internal organs have been preserved.

One example of an Eocene mammal is *Leptictidium*; a small, bipedal shrew-like animal, with a long tail for balance. It would have hopped along the forest leaf litter in search of food. It had a varied diet: small lizards, insects and plant matter have been preserved in its stomach contents. *Leptictidium*, at 20 cm tall, is typical of the small pioneer mammals that proliferated after the K/T event. The Eocene forests of Germany were teeming with life; the hubbub of small animals scrabbling in the leaf litter in search of food congregated as tumultuous forest floor communities. Above, in the forest canopy shrouded in the early morning mist, bird chorus clamoured aloft, while pond-dwelling frogs and toads introjected their ritualistic symphony. Such was the busy communities of small animals in the Eocene, during a time when the recovering Earth could no longer support huge animals.

This was no nirvana; these small Eocene mammals were by no means free from predators, of course. The dinosaurs left behind a legacy: giant, flightless birds terrorised the forests. In the absence of other large predators, giant birds were able to achieve some significance in the Eocene. *Gastornis*, at 1.5 m tall was able to prey on small ancestors of the horses in the humid European forests; seizing them with its huge, clawed foot, using its wings and tail feathers for balance. The victim was then dismembered with the huge, powerful beak, designed to tear flesh from a carcass. Terror birds of this kind lived predominantly in South America throughout the Tertiary and grew to heights of 3 m tall, so adapted to capture larger prey latterly.

As the Tertiary period progressed, mammalian faunas diversified greatly. During the Oligocene and Miocene epochs, Australia and South America had developed their own unique faunas. Marsupials were the main mammal group on the isolated continent, Australia. More varied and diverse forms were found here than any other places around the world. Many convergences exist between Australian marsupials and placental mammals from other parts of the world; as they both evolved to solve similar evolutionary problems. For example, the marsupial wolf, *Thylacinus*, looked remarkably like a dog or fox. Similar convergences exist in marsupial moles, anteaters, climbing insectivores, leaf-eaters and grazers. Convergent evolution arises when, in different localities, totally unrelated animals are presented with the same challenges, just by chance. As these animals adapt to the new discrete bilateral environments, modification with descent eventually produces new species that are perfectly adapted to the specific, but isolated environments. Being adapted to identical environments, or, more specifically, ecological niches within the environment, the two species will have very similar characteristics that are prerequisite for said niche, but importantly are derived from separate ancestry, so they have radically different genetics and could not breed with one another.

The most spectacular Australian marsupials date back from the Pliocene and Pleistocene. Herds of the giant wombat, *Diprotodon*, the size of a hippopotamus, dominated the landscape alongside the giant, short-faced kangaroo, *Procoptodon*. *Diprotodon* had heavy limbs and broad feet to support its weight, while *Procoptodon* hopped quickly, just as kangaroos do today at speeds of 45–55 km/h. Due to their size, these herbivores would probably be safe from predation, but the young and smaller species would have been hunted by *Thylacoleo*, the marsupial lion. Its heavy, 25 cm long skull had strong canine-like incisors, perfect for dispatching its prey. In summary, convergent evolution of marsupials in Australia produced populations that were analogous to those of placental mammals in other parts of the world, with stable food chains in conventional ecosystems.

Similar to Australia, South America was a continent isolated from other parts of the world for most of the Tertiary period. Like Australia, South America had a spectacular endemic (geographically restricted) population of unique faunas, evolving independently and often dissimilarly to mammals from other parts of the world. South America had its own group of marsupials that mimicked modern placental counterparts. The South American marsupials were less diverse and abundant than those in Australia and were mainly insectivorous, such as shrew-mimics, or carnivorous; being analogous to cats, sabre-tooth cats and dogs.

While South America had less abundant populations of marsupials, some more characteristic mammals were the xenarthral mammals, which included armadillos, tree sloths and anteaters. The name xenarthra means, "strange joints", which refers to the supplementary articulations between some of the vertebrae. Also, there was a peculiar arrangement in the hip girdle; the ischium and ilium were fused to the anterior caudal vertebrae. This group are also known as the edentates, which refers to another characteristic of the group: the lack of teeth. The armadillos (family Dasyprodidae) radiated in the Oligocene and Miocene epochs and, like the modern armadillos, had a bony head shield, partially-fixed body armour and a bony covering over the tail. The most spectacular armadillos were the glyptodonts from the Pliocene and the Pleistocene. They reached tremendous size, weighing up to two tonnes (400 kg of which was armour), and were suitably fortified against sabre-toothed marsupials. From the Miocene onwards, sloth evolution followed two main ecological lines: some, the modern tree sloths (family Bradypodidae), remained small and adapted to life in the trees, while others, the ground sloths (family Megatheridae) achieved great size. The largest of all was *Megatherium*; at 6 m tall it could rear up to feed on the leaves of trees; pulling branches towards itself with its long hooked claws.

Prehistoric South America, during the Tertiary, also had ungulate (bearing hooves) groups. These may be the relatives of modern horses, cows, rhinos, pigs, elephant etc., or perhaps independent groups altogether. These animals were the grazers that evolved in response to the newly established grasslands. Grasses evolved significantly in the Oligocene, when the development of prairies originated. Hitherto, small grassy areas were present in woodlands, but with the evolution of grazers, open grassland was created in regions where woody plants were kept in check by the grazers. Today, grassland covers 30% of the land surface on Earth, and, therefore, are a very important habitat. Many of the large mammals today are found in these regions; grasslands are a rich habitat that support large grazing animals as well as the predators that hunt them.

The unique faunas of South America underwent a radical reorganisation of communities about three million years ago, during the Pliocene and into the Pleistocene. The event that propagated it, which has become known as the Great American Interchange, is the opening of the Central American land bridge, which enabled populations of North American animals to move into South America and vice versa. This resulted in the complete unsettling of established ecosystems and extinctions of many species. Animals from both continents had new competitors and had to battle with the immigrants for the ecological niches to which the natives were no longer entitled; in natural selection, only the fittest survive and have the right to commandeer that particular position in the ecosystem.

North American mammals, including rabbits, dogs, horses, deer, bears, pumas and mastodons headed south, while the South American armadillos, glyptodonts, ground sloths, anteater and monkeys headed north. This interchange had complex ecological implications: the spontaneous arrival of new genera, resulting in competition and extinctions of weaker genera, changed the faunal constitution of the Americas; furthermore, other genera diversified by speciation. Paradoxically, mammal populations increased markedly after the land bridge appeared. This may be attributable to the possibility of some North American immigrants insinuating themselves, that is, found ecological niches without competition i.e. vacancies where South American species had not adapted to exploit the niches. Also, speciation may be more significant than outright extinction of weak genera, as animals changed their habits and adapted to new roles in the ecosystem in order to alleviate the intense competition. Again, this is another example of a quirk of evolution, where catastrophes, in fact, benefit biodiversity.

The main victims of the interchange were the South American ungulates and xanarthrans, most of which became extinct. Competition from northern invaders may not be the only cause of their demise, though. While the competing horses, deer and mastodons may have, in the long run, been superior, the replacement was not gradual. 26 genera of South American large herbivores decreased to 21 during the interchange, then abruptly rose to 26 again. The native genera, for some reason, were already starting to decline when the North American genera started to arrive. Also, the extinctions in the Pleistocene, which saw the loss of the glyptodonts and giant ground sloths cannot be explained by the invasion 2.5 million years before. It is probable that some other cause resulted in the decline of the unique South American genera. It could be that changing climate didn't suit the southern groups as well as the northern invaders, so the immigrants were not inhibited in any way by competition with the natives and by and large simply fitted into vacant niches.

As the Tertiary progressed, the placental mammals did indeed proliferate, becoming a very successful group. Most of the world was dominated by placental mammals, with only South America and Australia having their own unique faunas and by the Pliocene, placentals had invaded South America also. An explosive phase of radiation of placental mammals took place in North America, Asia and Europe during the Eocene, which saw the derivation of many, now extinct, groups, as well as the establishment of modern orders. Placentals continued to radiate and diversify throughout the Tertiary period to become the dominant group during the rise of mammals. The widespread success of placental mammals is considered to be an adaptive radiation: presumably, placentals had some key adaptations that proved advantageous and favoured their part in competition. These adaptations are speculative, but perhaps placentals proved superior having greater intelligence, extended parental care and improved dentition. Features such as these may have enabled placentals to capitalise on the vacant ecological niches present in the Eocene after the end of the dinosaurs and then again in the South American Interchange.

Perhaps the best example to illustrate the adaptive radiation of the placental mammals, at least in North America, is the Fort Union formation of the Crazy Mountain Basin, Montana. Typical of the Eocene Epoch, there are no large mammals present; very few exceed sheep-size. Most of the groups present in the fauna (about 75%) are now extinct. However, some of the modern orders are present and the rest appeared later in the Eocene.

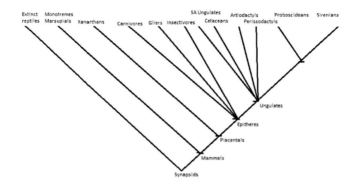

Figure 7: Cladogram of mammals.

In the six million years after the dinosaurs, the Eocene mammals had managed to achieve notable diversity. Many of the Eocene mammals were small and shrew-like or rabbit-like in appearance, generally feeding on insects, as with the example of *Leptictidium*. Early rooters and browsers had also evolved: *Phenacodus* had long primitive legs and is thought to be ancestral to ungulate horses. Larger mammals evolved later in the Eocene, such as the rhinoceros-like *Uintatherium*, which had bumps in the skull and, in males, 15 cm long canine teeth, presumably to serve as tusks for fighting. With the establishment of numerous herbivorous mammals, by the latter stage of the epoch, several carnivorous mammals had evolved to prey on them. Peculiarly, these meat-eaters were ungulates, called mesonychids; totally unrelated to modern carnivores. The largest, from the late Eocene of Mongolia, was *Andrewsarchus*. It grew to 5–6 m in length and had a massive skull: 83 cm long and 56 cm wide; the largest of any terrestrial carnivorous mammal. Animals of this kind became the apex predators of the time.

Many of the mammals at that time were insectivorous. Often regarded as the most primitive of the modern orders of placental mammals, the order insectivora arose in the Mesozoic. This

125

group includes the shrews, moles and hedgehogs. Shrews arose in the late Cretaceous, but are only known from jaw fragments. The moles and hedgehogs both arose in the Eocene. Close relatives of the insectivora, the archonta, include bats, tree shrews, flying lemurs and primates. The tree shrews and flying lemurs have specialised niches and, as a result, are much less abundant than either primates or bats. The bats (order chiroptera) include about one thousand species today; their success may be attributed to their advanced flying capabilities. There are two groups of bat: the megachiropterans, or fruit bats, and the more abundant insectivorous microchiropterans. Of the early bats, the best known is *Icaronycteris* from the early Eocene. The main microchiropteran features were already developed: the elongated fingers (not thumb) supported the wing membrane, the shoulder girdle is modified to hold large flight muscles on the expanded scapula and broad ribs and sternum. Strong hind limbs with feet turned backwards enabled it to hang upside down and the ears were adapted for echolocation.

A remarkably successful order of placental mammals are the rodents. With a total of 1,700 species today, rodents comprise 40% of living mammal species. Their success is testament to their adaptability, enabling them to coexist in a human landscape by changing their habits to benefit themselves. This same predisposition to modify habits, to utilise available resources meant that, in prehistoric times, rodents became a major group of mammals. Rodents are characterised by their teeth; having an upper and lower pair of deep-rooted incisors, which, unusually for mammals, grow continuously throughout life. The teeth only bear a veneer of enamel on the front face, so that the dentine behind wears faster, creating a sharp edge of enamel for gnawing wood, nuts and husks of fruit. The rodents first appeared about 40 Ma in the Eocene of Eurasia and North America. The largest group are the sciurognathous rodents, which include squirrels, dormice, beavers, hamsters, mice, rats, voles and some extinct groups. By the Oligocene and Miocene, the hystricognaths arose in Africa, which radiated to include living examples, such as guinea pigs, capybara, chinchillas and porcupines. Capybara are the largest living rodents; at 50 kg, they occupy an ecological niche more comparable with deer than a rat. However, they are dwarfed when compared to some giant hystricognaths of the past, with some genera growing as large as a rhinoceros.

At the beginning of the Tertiary, the groups of smaller mammals mentioned above were predominantly the main groups and the large browsers and grazers had not evolved yet, but their ancestors did indeed exist alongside the rodents and insectivores. The early ungulates, albeit small in size compared to their descendants, radiated in the early Tertiary into the two recognisable groups: the perissodactyls (the ungulates with an odd number of toes), including horses, tapirs and rhinoceroses; the artiodactyls (ungulates with an even number of toes), such as pigs, cattle, sheep, deer and camels.

Among the first perissodactyls were the horses, albeit miniatures, the size of dogs, such as *Hyracotherium* from the Eocene. Horses have undergone major changes during the course of their evolution. For example, early forms, like *Hyracotherium*, had four front toes and three on the back foot, while the modern horse has one hoofed toe on each foot. The teeth also changed, in accordance with the animal's diet. *Hyracotherium* had small leaf-crushing molars, while the modern horse has deeper molar for grinding course silica grass. The changes in teeth and limb structure are linked to the increase in overall body size of horses over the course of their evolution. These changes have been attributed to environmental changes taking place in the early Miocene, when grasslands began to spread to North America. Early horses, akin to *Hyracotherium,* were browsers, feeding on the leaves of small bushes and trees. As the humid forests retreated, and grasslands spread, new horse lineages appeared. While *Hyracotherium* relied on camouflage to evade predators in the forest, this was inadequate in open grassland; horses grew larger, with long limbs and a single hoof; becoming adapted to run at great speed to escape predators.

Other living perissodactyls, and indeed, probably relatives of the horses, were the tapirs of Central and South America and Southeast Asia, and the rhinoceros of Africa and India. The early tapirs looked similar to the early horses and radiated after the Eocene, where the major evolutionary changes that distinguished them conspicuously from horses was the development of the short trunk. The rhinoceroses had a much more diverse history. Similar to the early horses and tapirs, early rhinoceroses, such as

Hyracodon, were medium-sized running animals, which lacked horns. At the extreme, the Oligocene rhinoceros *Indricotherium* grew to enormous size; it is the biggest land mammal of all time: 4–5 m tall at the shoulder and probably weighed about 30 tonnes (the same as five African elephants). The horned rhinoceroses began to radiate in the Miocene. By the Pliocene, they became extinct in North America, but ancient lineages, such as the woolly rhino, *Coelodonta*, remained in Europe to the end of the Pleistocene.

Two extinct groups of perissodactyls that were probably ecologically analogous to the living families are the brontotheres and the chalicotheres. Brontotheres, such as *Brontotops* from the USA, radiated in the Oligocene. They had a Y-shaped structure at the end of the snout, like a horn, that was probably covered with skin in life and may have functioned as a sexual display structure. *Brontotops* was 2.5 m tall at the shoulder and heavily built, but this group was replaced by large rhinoceroses. The peculiar-looking chalicotheres persisted longer than the brontotheres, lasting until the Pleistocene. *Chalicotherium* looked horse-like, except they had the posture of a gorilla. Having long forelimbs and short hind limbs, they may have been able to erect themselves in a bipedal manner to grasp the foliage of low branches. Claws on the back feet may have been for digging, while the fingers had hooves. *Chalicotherium* would have walked with its fingers curled, bearing its weight on the knuckles, as gorillas do. The hooves were protected from the hardship of walking, maybe because they were important feeding apparatus.

Living alongside the perissodactyls in the Tertiary were the even-toed artiodactyls, which themselves may be divided into two groups: the first includes cattle, deer, sheep, antelope, camels and giraffes; the other pigs and hippopotamus. The first artiodactyl radiation; the main group of selenodonts were the ruminants: cattle, sheep, deer and antelope, which occurred in the Oligocene and Miocene. This method of digestion involves predominantly consuming grass, which enters the rumen where it is partially digested by bacteria. The food is then regurgitated for rumination, or "chewing the cud", before being processed through other stomachs. In this way a cow, which has four stomachs, can maximise assimilation of nutrients.

Pigs and hippos, also known as the bunodontia, never achieved the same diversity as the selenodonts. However, in the early Oligocene, large areas of North America were populated by giant pig-like animals, called entelodonts. Entelodonts would have frequented ponds and lakes, feeding on the wide range of plants nearby and, probably like pigs today, were omnivorous and would scavenge the carcass of a dead animal. Entelodonts grew to 2–3 m in length and some genera had long canines: tusks to fend off rivals during the intense competition that takes place in highly localised territories. These animals would have behaved as reprobates, fighting and tussling over the right to frequent a particular pond area.

Elephants (order proboscidae) today include only two species, the African and the Indian elephants, but this is by no means representative of their former diversity. Their early evolution took place mainly in Africa, with examples such as *Moeritherium* from the late Eocene, which had a relatively long body and was 1 m tall; probably living its life in fresh water areas, along with the entelodonts, like a hippo. It had the second upper and lower incisors enlarged as a short projecting tusk, which seems to be an important ornament for these high-competition localities.

As elephant evolution continued through the Oligocene and Miocene, several lineages of mastodon appeared, which culminated in modern elephants. The mastodons tended towards larger size and the head became heavier, not least because of the large tusks, which could only be supported by a short neck. As a result, elephants couldn't reach the ground with their mouth; hence, the trunk of proboscideans became longer. The tendency towards larger size may have been a response to oppose predation from large carnivorous genera that existed at the time. As the mastodons grew larger, they required more food and intraspecific competition within and between herds would have increased, so the development of tusks would have been necessary to fend off rivals during conflicts thereby instate dominance hierarchy. The tusks also functioned in feeding, used to dig up roots and tubers, scrape bark from trees and push over trees. With the increased appetite and movement onto savannas, sourcing more abrasive food, not to mention the longer lives of more modern elephants of up to seventy-five years, the issue of tooth wear became a major problem. *Moeritherium*, like other mammals, had all six pairs of cheek teeth operating at the same time. Modern elephants, by contrast, only have one pair in place at a time. The final pair lasts the elephant to the end of its life, when the tooth becomes worn down and the animal can no longer feed.

A diversity of proboscidean families lived during the Tertiary, from the Miocene to the Pleistocene. The deinotheres, such as *Deinotherium*, had a pair of tusks curling under the lower jaw, probably for scraping bark from trees. The gomphotheres, as another example, had four short tusks formed from extensions of the upper and lower jaw. Perhaps the most charismatic extinct proboscidean is the woolly mammoth, with its thick coat and large tusks to brush away the snow from the plants on which it fed in the Pleistocene tundra. The proboscideans were once widespread, found in Africa, Europe, Asia, North America and even some families in South America.

As the herbivorous mammals adapted and diversified to fit the variety of ecological niches that developed throughout the Tertiary, so did their predators. Predators adapted in response to the prey opportunities that derived from the broad diversification of mammalian herbivores. The carnivores comprise their own taxonomic order and are classified into two main groups: the feliforms, which includes cats, hyenas and mongooses, and the caniforms, which includes the dogs and the arctoids (bears, racoons, weasels and seals). Cats have evolved dagger-like teeth several times independently throughout their history. While modern lions, tigers and some extinct forms, such as *Deinofelis*, have canines for puncturing flesh, some extinct cats had massive sabre-teeth, up to 15 cm long. Among these were the marsupial cats, *Thylacoleo* of South America and *Smilodon* of North America and Europe; both totally unrelated, yet had evolved the same predatory characteristics to takedown large prey. The lower jaw can be dropped very low to enable the sabre-teeth, which were flattened and backwards-curved like a knife, to be inserted. It may be expected that specialised killing apparatus of this kind was suited to piercing the pulmonary vein in the thick necks of large herbivores, but the purpose of the sabre-teeth was to tear chunks of flesh from the prey, rather than to inflict punctures, causing the victim to bleed to death. The infamous sabre-teeth acted more like carving knives to dismember the meal, rather than as severe weapons. One can imagine a pack of sabre-tooth cats chasing a terrified antelope towards a corral, since the heavily-built cats were not adapted for open pursuits like a cheetah, and leaping onto the unfortunate creature. Having brought it down, the *Smilodon* began to pull the carcass to pieces with their long teeth.

Lastly, in addition to colonising the land, mammals also took to the water. The whales (order cetacea) evolved from terrestrial origins; embracing the adaptations necessary for life at sea. The first known whale, *Pakicetus* from the Eocene of Pakistan, was a semi-aquatic predator that could still move on land, dwelling on coastlines, as walrus do. By the Oligocene, whales became fully aquatic and some grew very large, such as the 20 m long *Basilosaurus.* From this point onwards, whale evolution followed its present course, with cetaceans being divided into two groups: the toothed whales, such as dolphins, porpoises and sperm whales, and the baleen whales, such as the humpback and blue whale.

With the dinosaurs gone, at the start of the Tertiary mammals were free to rise to dominance. Minimal predation initially and limited competition for resources propagated a boom in evolution that saw the rise of mammals to became the dominant group. Different continents developed their own unique faunas. Perhaps the most important development for the mammals was the origin of grasslands because this became a very important ecosystem as the mammals readily adapted to feed on grasses and predators, in turn, adapted to hunt the grassland herbivores. The grassland herbivores varied from large herds of ungulates to heavy animals, such as elephants, which at this stage of mammal evolution were a very important group. In localised areas, such as at ponds, diverse communities flourished.

The Ice Age

The thriving and diverse mammal populations that have arisen so far in the Tertiary period, by the Pleistocene epoch, became subjected to challenges which occurred as a result of the changing climate. The Pleistocene epoch, spanning a couple of million years, is marked by five or so major ice ages, where the

North Pole advanced southward, covering much of North America and Europe in ice sheets. During this period, the Earth experienced major climatic changes, contributed towards by both astronomical and Earth-bound factors. The ice ages were separated by inter-glacial episodes of warmer weather; the last of which started 11,000 years ago at the end of the last ice age.

The opening of the Central American land bridge in the Miocene that enabled the passage of animals between the North and South American continents, blocked a major equatorial current that flowed from the Atlantic to the Pacific; deflecting water northwards into the Atlantic in a much strengthened Gulf Stream. This brought warmer water, more evaporation and more precipitation into the North Atlantic. The Northern Hemisphere was now primed for a glaciation event in the Pliocene when a drop in solar radiation and/or volcanism triggered the formation of major ice sheets. Milankovitch Cycles reached a phase in the Earth orbit that promoted warm wet winters; increased snowfall over North America, Greenland and Northwest Europe accumulated in thicker and wider snowfields. Reflection of the sun's radiation off the snow back into space, through a process known as the albedo effect, reduced the intensity of the rays on the ground, allowing the ice sheets to spread across the Northern Hemisphere into a glacial landmass. The Earth was gripped in the Ice Age.

The rapid onset of glaciation initially impacted on the Pliocene mammal faunas, but the Ice Age mammals of the Pleistocene, oddly, were not affected by the rapidly fluctuating climate as the ice sheets regressed and advanced. The changing climate, although rapid on a geological timescale was slow enough to allow animals to migrate north and south with the changeable ice sheets. Changes in sea level accompanied every glacial advance and retreat. Each major glaciation caused a drop in global sea level as water was stored in the ice; lowland areas became extruded and joined major landmasses together, which facilitated migrations. Alaska and Siberia were joined across what is now the Bering Strait and Greenland joined North America in one giant northern continent, creating vast areas of boreal forest and tundra in which ice age animals lived.

North America had mastodons, mammoths, giant bison, ground sloths and sabre-tooth cats. Eurasia had most of these as well as woolly rhino and giant deer. All these species are now extinct. The Pleistocene megafauna had become adapted to the glacial climate; some were ice age specialists, such as the woolly mammoth and woolly rhino. Large body size, long coarse hair with an insulating finer layer beneath and a thick layer of fat meant that they could keep warm, while having slow metabolic rates, which enabled them to live on the poor-quality tundra vegetation, such as sedges, grasses which they foraged for, clearing the snow with their tusks and horns, as well as browsing tundra trees, like alder and birch. The only changes made to their anatomy over episodic periods of warmer interglacials was body size, with Siberian mammoths being about 20% larger than in the coldest times. Indeed, the animals proliferated during the Pleistocene ice ages, while animals that existed in more temperate climates existed in smaller, but stable populations. Giant ground sloths and glyptodonts were restricted in habitat, but lived unpressured by climate. The last ice retreat, about 11,000 years ago, was not climatically unique; the megafauna survived the past few glaciations, so why did they die out at the beginning of recent times?

Indeed, the megafauna, large mammals as well as giant birds were effected most and these extinct animals were never replaced. The extinctions occurred in different regions at very different times, but were all sudden and closely linked in time with the arrival of humans. Few fossil remains of megafauna appear alongside human remains, which implies coexistence was brief. Twenty genera of large Pleistocene mammals became extinct during the two million years before the last ice age, yet 33 genera were lost in less than 3,000 years at the last glacial regression. Evidence suggests that human hunting was the cause.

Humans travelled into North America from Siberia across the windswept land bridge, when the Bering Strait was a dry land area, Beringia. As the last ice age regressed, the ice sheets of the Canadian Shield and Rocky Mountains retreated; the ice barrier opened, enabling the skilled hunters from a distinctive weaponised culture to pass into the Americas. Remains of mammoths and mastodons have been found with cut marks and tool points; direct evidence of slaughter and butchering of the animals and bones assembled together is evident of large chunks of meat cached for winter; stored in shallow water under a thin sheet of ice, as Inuit do today. Bones and tusks were also used to make tents by the modern humans. The megafauna were the most susceptible to human predation because they move relatively slowly and can't hide, so are easy targets. They exist in small populations and breed slowly, so extinction through overhunting was inevitable.

During this period, mammalian faunas underwent major changes; likely as a consequence of the changing climate, but also from a new threat: man. These combined factors caused extinctions to occur, with a generalised decrease in biodiversity. In South America, 46 genera (80%) died out. Most effected were the megafauna: the giant mammals. In North America, 73% of the large mammals (33 genera) including mammoth, mastodons, glyptodonts and ground sloths died out. Australia was also effected, but Africa and Asia less so. What is certain, the arrival of man had a profound effect on the giants of the Ice Age and was destined to significantly change the world, starting a new chapter in the history of life on Earth.

Human Evolution
The Bipedal Ape

Humans are one of two hundred living species of primate. Primates are characterised by a number of features that are associated primarily with three major adaptations: agility in the trees, acute brain and eyesight and parental care. Typically living in trees, primates have anatomical features that enable them to exist in this habitat. Among these adaptations are a very mobile shoulder joint, which can be rotated in a complete circle, grasping hands and feet in which the thumb and big toe may be opposable, flat nails rather than claws, and sensitive pads on the digits.

Relative to their body size, primates have larger brains than other mammals, enabling them to cope with the complexity of the forest ecosystem. Also, their eyes are large and close together on the front of the face and the snout is reduced. The flattened face enables them to look forward with a large overlap in the fields of vision of each eye, allowing them to have stereoscopic, or three-dimensional sight. This is useful for judging distances when jumping between branches. The enhanced brain capabilities may also be linked with the extended parental care displayed by primates, which is necessary to teach the offspring the complexities of living in the forest.

Primates are well-known from the Eocene onwards. Major stages in ape evolution, which ultimately led to us humans, took place in the Miocene of East Africa. The apes (superfamily hominoidea, which today include gibbons, orangutans of Asia, the gorilla and chimpanzee of Africa and humans) made a significant proportion of the mammalian fauna. These early apes seemed to have both climbed trees and lived on the ground. They would have walked quadrupedal on the ground and along thick branches of trees, feeding on tough leaves or fruit. Some, such as *Sivapithecus* were the ancestors of orangutans. A relative of *Sivapithecus* was *Gigantopithecus*: a monstrous animal at 2.5 m tall, 270 kg and is thought to be the inspiration of Yeti stories in Asia and Bigfoot in North America. It died out as recent as maybe 250,000 years ago, and would have survived long enough to have been encountered by the first humans.

The line of great apes that led to human ancestors, independently of *Sivapithecus,* were the first bipedal apes: the australopithecines. Evidence of bipedalism among these apes comes from 3.5 million-year-old footprints. A carbonite eruption on the plains near Laetoli, in northern Tanzania, during the late Pliocene, produced ash deposits, which, after the first rain dried, set as a cement, preserving any impressions made into it as well as the unfortunate animals that fell victim to the ash. The ash fall on the barren plains, scattered with acacia thorn trees, buried rodents and hares in their burrows. Repeated ash falls ensued afterwards. Rain wetted the surface of the ash, leaving footprints; trails of African animals were preserved that illustrate the movement of elephants, giraffes, antelopes, ostriches, hyenas, as well as extinct examples, such as chalicotheres, onto the plains for the onset of the rainy season, just as they do today. Most remarkable was the 25 m trail of three *Australopithecus*; a family group, inferred from the size of the footprints, being small, medium-sized and large. The middle *Australopithecus* walked precisely in the footprints of the large, while the tracks of the small infant follow closely alongside, interpreted as holding hands with the mother as the group trudged through the deep ash slurry. Crucially, this is evidence of apes walking bipedally, before evolving towards vaguely human characteristics.

Walking fully erect would have been advantageous to the australopithecines, enabling them to walk further, foraging over greater distances, following the African migrations to where food was more plentiful. The arms were free to carry feeble infants along the journey and they could see further across the plains to identify and evade threats. It could be that bipedal walking in apes evolved, by extension, as the forelimbs of primates predominantly were used for gathering food, while the hind limbs tended to be for locomotion. Primates would stand in trees and collect fruit with their hands. If this behaviour directed towards an erect posture, this may have transferred to standing upright on the ground. This would have made walking easier, so the bipedal apes could travel more readily to regions where food was more plentiful, thus they were at a selective advantage. Examining the skeletons of australopithecines reveals that the first (uppermost) rib is jointed to only one vertebra. Humans and kangaroos are the only living mammals to have this feature (other mammals have the ribs jointed between two vertebrae) which is associated with upright stance, perhaps relating to changes in the shoulder, or to freer respiration. This suggests that all australopithecines were bipedal. They moved much in the same way, waddling across the ground, or for a lot of the time, climbing in the trees. The australopithecines were not yet the most efficient walkers, but evolution towards the first people was directed by the success of bipedal locomotion.

Hominids: The First People?

Paleoanthropologists have deduced that the line to modern humans includes at least six hominid species; three are australopithecines and three homo. The earliest hominid is *Australopithecus afarensis*, whose three-million-year-old remains were discovered in the early 1970s in Ethiopia, and nicknamed Lucy. Lucy is 1.2 m tall and 40% of the bones were present, so the skeleton was well-complete. Many primate features were present: a small braincase of only 400 cm^3, long arms and short legs, with curved digits, implying that Lucy was still living in trees. However, the hind limbs and pelvis were fully adapted for bipedal locomotion. Similarly to ourselves, australopithecines were built for trotting endurance, rather than sprinting; helpful for foraging over large distances across savanna. However, susceptibility to predation from speedy carnivores, such as big cats, may have been an omnipotent jeopardy in the open plains, but the long arms of *Australopithecus* would have had sensitive motor control; well-coordinated enough to throw stones and swing heavy logs, bombarding a predator in group defence strategy, as troops of baboons do today. The major difference in the pelvis of Lucy, when compared with that of a modern woman, is that the pelvic canal is broad, but narrow from front to back. In *Homo*, by contrast, the birth canal is much rounder to be able to expel a baby with a much larger head and brain. It could be that broader shoulders on australopithecine babies is the cause of this differentiation.

The ecological role of australopithecines may have been analogous to today's baboons. They are likely to have sheltered in high places, such as trees or rock faces, and foraged among open savanna in troops, with a cohesive social structure. Group defence would have been an effective protection from predators, offering security to the australopithecines, who were vulnerable individually. Being bipedal, australopithecines would probably have been ecologically superior to baboons; the arms were free to gather food, which could be collected and taken away, reducing the time spent out on risky expeditions. Fossils of australopithecines show that they often lived around lake edges. In addition to the fibrous plant material, their staple diet, preserved crab claws, turtle and crocodile eggs suggest that a variety of proteinaceous foodstuff were available to them. The transition to protein-rich foods may have facilitated the growth of brains in the australopithecines.

Australopithecines continued to live in Africa for a further two million years. Later forms evolved, which had advances over *A. afarensis*. They were larger (1.75 m tall, 70 kg in weight) and had a larger brain capacity of 550 cm^3. These include *A. robustus* and *A. boisei*, which were better adapted bipedally. However, many of these deviate from the human lineage. The previously mentioned examples are classified in a group known as robust australopithecines. They had a more projecting face and large molars for grinding tough plant matter. A ridge along the top of the skull is present for the attachment of strong jaw muscles, needed to chew this fibrous vegetarian diet. They were ecologically similar to gorillas and chimpanzees rather than humans. Our own lineage had much reduced teeth and jaws, a flatter face and were more lightly built. Most important was a larger brain case. As australopithecines got a taste for nutritious meat, improved intelligence enabled active hunting and the use of tools. Technological advances accompanied the appearance of our own genus, *Homo*.

The oldest species of our genus is *Homo habilis*, from Kenya. It had a large brain (650–800 cm^3). This brain size, for a

1.3 m tall hominid comes very close to the modern human range. It had dexterous hands, which had the manipulative ability to fashion tools; hence, the meaning if its name, "handy man". *H. habilis* has been found alongside remains of various Australopithecine species, promoting the idea of several human species living, and probably interacting together 1.5 million years ago.

H. habilis fashioned clumsy tools by hammering rocks together to produce flakes used for cutting and scraping. An excavation in the Turkana Basin revealed a hippopotamus close to an ancient riverbed. Stones from the gravel bank were used to make butchery tools. Marks on the bones show that they had been scraped and the tendons and ligaments cut to remove joints of meat from the carcass. No evidence was found to suggest the hippo was killed using tools. The use of tools demonstrates a certain level of intelligence in the early *Homos*. Furthermore, some palaeontologists advocate a higher level of sophistication in the tool maker than previously realised. Reconstruction and use of these tools reveals how they needed to be made from carefully selected stones and are well-designed for effectively cutting hides, butchery and breaking bones. Moreover, examination of the tools could even determine that *H. habilis* was right-handed. Tools were made on site from suitable stones and these sites were revisited many times in intervals over fifteen years, which indicates an intelligent return to useful sites for processing carcasses.

A more graceful gait than in australopithecines meant that *H. habilis* could proficiently roam open country in search of a fresh carcass to scavenge. The rudimentary tools were suitably effective to cut open the thick hides of large animals, difficult for vultures and hyenas to pierce. Carcasses of smaller animals would be rare to come by because these scavengers would readily devour them. Nevertheless, evidence of cut marks have been found on the remains of small to medium-sized prey animals, which might indicate active hunting by *H. habilis*. It is reasonable to assume that *H. habilis* learned to locate large carcasses, for example by following vultures, or observe where leopards cached their meals in trees. With the use of tools to help in processing meaty meals, the first people derived their own ecological role. To cooperate effectively in finding food, it could be that primitive means of communication, such as hand signalling, were required within social groups.

Modern Humans

Around one million years ago, humans spread out of southern and eastern Africa. Hitherto, all human evolution had occurred on this continent, but now pioneering species begin to spread elsewhere across the globe. Derived in Africa about 1.6 million years ago, *Homo erectus* was the first widespread human, having certain advances over *H. habilis*, which enabled it to expand its population. *H. erectus* stood 1.6 m tall and had a brain capacity of 850 cm^3, although the skull was still primitive in morphology, with a thick eyebrow ridge and a broad jaw with no distinguishable chin. However, the rest of the skeleton is more modern and absolutely bipedal, perhaps enabling the vast migration of *H. erectus*. Large hip and back joints were designed to take the stresses of full running, while all special adaptations for climbing were omitted. It became widespread in China 0.6–0.2 Ma, with 40 individuals found in Zoukeouton cave, which had evolved larger brains of 900–1100 cm^3. With this came cultural advances: living in a tribe in semi-permanent homes and hunting cooperatively.

The superiority of *H. erectus* over *H. habilis* can be inferred from technological advances in the tools used. *H. erectus* designed a wider range of tools, which required much more strength and precision to craft. The familiar flakes used to make cutting and scraping tools were present, while the stone cores were afterwards fashioned into axes and cleavers. With the development of these heavy duty tools, it is suspected that active hunting was a predominant occupation of *H. erectus* and it is tempting to correlate this with the extinction of a number of species at about 1 Ma. Among the extinctions, including sabre-tooth cats and some of their prey items, were the out-competed australopithecines and inferior *H. habilis*.

By this time, humans moved into Asia. Fossils of "Peking Man"; specimens of *H. erectus* from Java, Indonesia and later occupied caves outside Beijing, show advances over the African races. Brain size had increased to 1100 cm^3, associated with superior intelligence and they found solutions to living in a challenging seasonal climate. They were skilled hunters of deer and built fires. The success of *H. erectus* in Asia is indicated by their long-term inhabitancy of China. With greater intelligence, the social structure of their primitive culture would have become more complex. Perhaps, as a group of *H. erectus* sheltered in a cave from harsh winter weather, assembling a fire to generate warmth the first conversations developed. This would have started the novel process of communicating information and knowledge directly between individuals and enhanced learning between members of the group. This would be better than the abstract methods of taking and showing, or copying because more information is retained and understood by individuals. However there is no evidence indicating that *H. erectus* could talk; only inferences from social speculations.

Evolution had advanced the genus further by the time that *Homo sapiens* moved into Europe. Neanderthal man, regarded as being dim-witted, actually had a larger brain capacity than modern humans (1400 cm^3 compared to 1360 cm^3 on average) but was less functional. Neanderthals date back 100,000–35,000 years: his robust and compact form enabled him to thrive during the Ice Age. They would have differed markedly in appearance from modern humans, having big noses for breathing in cold air, large front teeth and no chin. The big incisors are thought to be an adaptation derived from manipulating animal hides with the teeth. The robust skeleton may be attributed to a lifestyle that required great physical strength. Stone-carved spear points are an example of the tools used; hunting with a spear would need strength to impale a large animal with a thick ice age hide. Enlightening evidence of a more developed social structure comes from ceremonial burials. Analysis of pollen samples from a grave of a Neanderthal man in Kurdistan, Iraq, show that he was laid to rest with carefully selected, decorative medicinal herbs with colourful flowers, belonging to seven species that all bloomed together in the spring. This suggests contemplation of the abstract world; a peculiarity of us humans.

Far from being a separate species, Neanderthals can be considered as a peripheral race of archaic humans that evolved from the same lineage as modern humans, perhaps descending from *H.erectus*, as a pioneering set of people that resided along the edge of the northern ice sheets of Europe and central Asia, adapting to the cold climate. Archaic humans, meanwhile, lived in warmer climate zones, so competition between the two races was minimal and neither can be regarded as being superior, being adapted to different environments. In transitional areas between the climatic zones, archaic humans and Neanderthals interacted and seasonally inhabited the same caves. The distinguishable features between them are less discrete in these regions as the strong Neanderthal features become more subtle in the warmer climates. There is even speculation of interbreeding, which has intriguing implications for genetics in modern human populations.

About 35,000 years ago, Neanderthals as a defined race disappeared. This coincides with the advances of modern humans. Cro-Magnons of Europe soon populated many territories around Eurasia. The Neanderthals were quickly replaced being a subordinate race to the intellectually superior Cro-Magnons, who crafted more advanced tools, such as projectiles; arrows were tipped with stone, as were durable bone-made spears. The successful hunting strategies of Cro-Magnons caused the simplistic Neanderthals to be usurped, the food was taken by the overwhelming Cro-Magnons, and the populations of Neanderthals dwindled. Indeed, the proficient hunting by Cro-Magnons had the capacity for wide-scale destruction; as populations grew, overhunting led to the extinction of many Ice Age species. The drawings of these animals on the cave walls were done by the last people to see these magnificent beasts alive.

It is clear that many races of modern humans, descended from *H.erectus,* became adapted to living in a variety of climatic zones and the species *Homo sapiens* became the dominant group in this current chapter of the history of life on Earth. Man has shaped the world to suit his needs. The development of agriculture arose to feed populations that could not be supported by natural ecosystems and vast settlements were built to house the populations. Advances in technology coincided with the frontiers of intellectual ability; the intelligence that humans evolved as a unique selective advantage continues to be the crucial adaptation to determine our success. Abstract thought and ingenuity enables us to combat the challenges that we are confronted with. Medicine, travel, food, exploiting natural resources are all technological developments that enrich our lives and have indeed become crucial to our survival.

Our intelligence and curiosity has prompted us to question our origin. Through scientific rigour, we have deduced an inkling into the 3.5 billion year story of evolution on our planet. Any rational scientific thought objects to the notion of creation by a divine deity, but when considering the complexity of the enigmatic phenomenon of evolution, any ridicule of religious philosophy seems unfair. Any derivation of unified reasoning to rationalise this axiomatic property of nature continues to be illusive. Perhaps at its most fundamental, at the core of evolution quantifiable physical and chemical properties govern the history, and indeed the future of life on Earth.